WHEN
THE
EARTH
HAD
TWO
MOONS

ERIK ASPHAUG

CUSTOM
HOUSE

HarperCollins books may be purchased for educational, business, or sales promotional use. For information, please email the Special Markets Department at SPsales@harpercollins.com.

FIRST EDITION

Designed by Bonni Leon-Berman

Title page art © Vadim Sadovski / Shutterstock

Library of Congress Cataloging-in-Publication Data has been applied for.

ISBN 978-0-06-265792-3

19 20 21 22 23   LSC   10 9 8 7 6 5 4 3 2 1

WHEN
THE
EARTH
HAD
TWO
MOONS

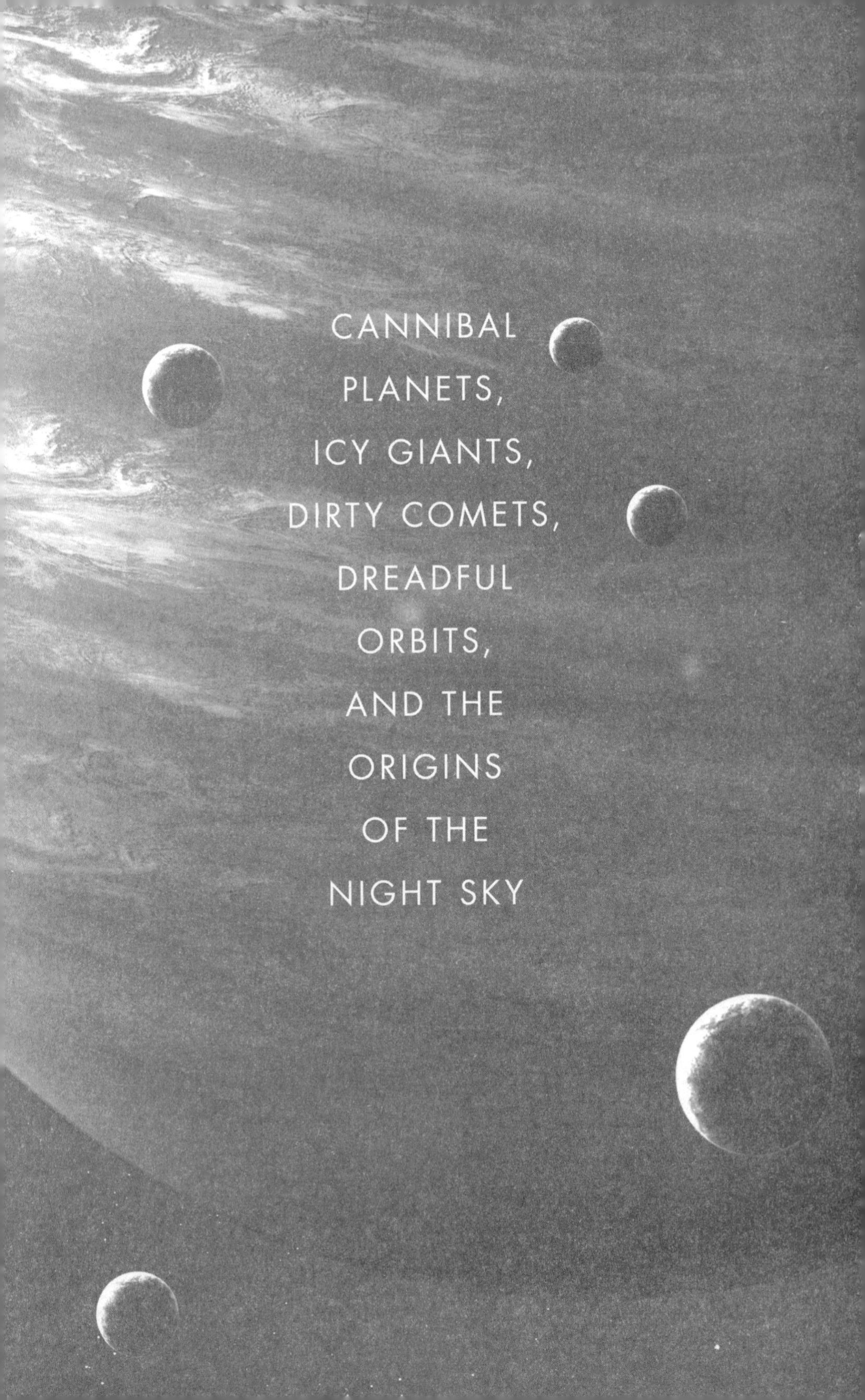

CANNIBAL
PLANETS,
ICY GIANTS,
DIRTY COMETS,
DREADFUL
ORBITS,
AND THE
ORIGINS
OF THE
NIGHT SKY

To Henry, Galen, and Phoebe

CONTENTS

A SHORT LIST OF
PLANETS AND MOONS

There are at least nine planets in the solar system (depending on who's counting) and they have almost two hundred known moons (natural satellites). Below are some of the most interesting and important ones.[1] Because some of the moons are oddly shaped, and the fast-rotating planets are oblate, what's given is the average diameter. Orbital distances of planets are in AU, where 1 AU is the Earth's average distance from the Sun, 149.6 million kilometers. The orbital distances of satellites are given in units of their planetary radius.

MERCURY
Distance from Sun: 0.39 AU
Diameter: 4,878 km
Mass: 3.301×10^{23} kg
Orbital period around the Sun: 0.24 years / 88 days
Spin period: 58.6 days

VENUS
Distance from Sun: 0.72 AU
Diameter: 12,104 km
Mass: 4.867×10^{24} kg
Orbital period around the Sun: 0.62 years / 226 days
Spin period: 243 days (retrograde)

EARTH
Distance from Sun: 1 AU (defined)
Diameter: 12,742 km
Mass: 5.972×10^{24} kg
Orbital period around the Sun: 1 year / 365.26 days
Spin period: 23.93 hours (sidereal day)

Moon
Distance from planet: 60.3 Earth radii
Diameter: 3,474 km
Mass: 7.35×10^{22} kg
Orbital period around Earth: 27.3 days (sidereal month)

MARS
Distance from Sun: 1.52 AU
Diameter: 6,779 km
Mass: 6.417×10^{23} kg
Orbital period around the Sun: 1.88 years
Spin period: 24.6 hours

Phobos
Distance from planet: 2.8 Mars radii
Diameter: 22 km

Mass: 10.8×10^{15} kg
Orbital period around Mars: 7.7
   hours

Deimos
Distance from planet: 7.0 Mars radii
Diameter: 12 km
Mass: 1.48×10^{15} kg
Orbital period around Mars: 30.3
   hours

JUPITER
Distance from Sun: 5.2 AU
Diameter: 139,822 km
Mass: 1.898×10^{27} kg
Orbital period around the Sun: 11.86
   years
Spin period: 9.9 hours

Io
Distance from planet: 6.03 Jupiter
   radii
Diameter: 3,643 km
Mass: 8.93×10^{22} kg
Orbital period around Jupiter:
   1.8 days

Europa
Distance from planet: 9.59 Jupiter
   radii
Diameter: 3,130 km
Mass: 4.79×10^{15} kg
Orbital period around Jupiter:
   3.6 days

Ganymede
Distance from planet: 15.30 Jupiter
   radii
Diameter: 5,268 km
Mass: 1.48×10^{23} kg
Orbital period around Jupiter: 7.2
   days

Callisto
Distance from planet: 26.93 Jupiter
   radii
Diameter: 4,806 km

Mass: 1.08×10^{23} kg
Orbital period around Jupiter: 16.7
   days

SATURN
Distance from Sun: 9.6 AU
Diameter: 116,464 km
Mass: 5.683×10^{26} kg
Orbital period around the Sun: 29.44
   years
Spin period: 10.7 hours

Mimas
Distance from planet: 3.18 Saturn
   radii
Diameter: 398 km
Mass: 3.75×10^{19} kg
Orbital period around Saturn: 0.942
   days

Enceladus
Distance from planet: 4.09 Saturn
   radii
Diameter: 504 km
Mass: 1.08×10^{20} kg
Orbital period around Saturn: 1.37
   days

Tethys
Distance from planet: 5.06 Saturn
   radii
Diameter: 1,072 km
Mass: 6.17×10^{20} kg
Orbital period around Saturn: 1.89
   days

Dione
Distance from planet: 6.48 Saturn
   radii
Diameter: 1,125 km
Mass: 1.10×10^{21} kg
Orbital period around Saturn: 2.74
   days

Rhea
Distance from planet: 9.05 Saturn
   radii

Diameter: 1,528 km
Mass: 2.31×10^{21} kg
Orbital period around Saturn: 4.52
days

Titan
Distance from planet: 21.0 Saturn
radii
Diameter: 5,150 km
Mass: 1.34×10^{23} kg
Orbital period around Saturn: 15.9
days

Hyperion
Distance from planet: 25.7 Saturn
radii
Diameter: 270 km
Mass: 1.08×10^{19} kg
Orbital period around Saturn: 21.3
days

Iapetus
Distance from planet: 61.1 Saturn radii
Diameter: 1,469 km
Mass: 1.81×10^{21} kg
Orbital period around Saturn: 79.3
days

URANUS
Distance from Sun: 19.2 AU
Diameter: 51,26 km
Mass: 8.681×10^{25} kg
Orbital period around the Sun: 84.02
years
Spin period: 17.2 hours (retrograde)

Miranda
Distance from planet: 5.08 Uranus
radii
Diameter: 472 km
Mass: 6.59×10^{19} kg
Orbital period around Uranus: 1.41
days

Ariel
Distance from planet: 7.47 Uranus
radii

Diameter: 1,160 km
Mass: 1.3×10^{21} kg
Orbital period around Uranus: 2.52
days

Umbriel
Distance from planet: 10.4 Uranus
radii
Diameter: 1,170 km
Mass: 1.17×10^{21} kg
Orbital period around Uranus: 4.14
days

Titania
Distance from planet: 17.1 Uranus
radii
Diameter: 1,577 km
Mass: 3.53×10^{21} kg
Orbital period around Uranus: 8.71
days

Oberon
Distance from planet: 22.8 Uranus
radii
Diameter: 1,523 km
Mass: 3.03×10^{21} kg
Orbital period around Uranus: 13.5
days

NEPTUNE
Distance from Sun: 30.0 AU
Diameter: 49,244 km
Mass: 1.024×10^{26} kg
Orbital period around the Sun: 165 years
Spin period: 16.11 hours

Proteus
Distance from planet: 3.77 Neptune
radii
Diameter: 420 km
Mass: 4.4×10^{19} kg
Orbital period around Neptune: 1.1
days

Triton
Distance from planet: 14.4 Neptune
radii

Diameter: 1,682 km
Mass: 2.14×10^{22} kg
Orbital period around Neptune: 5.9
  days

Nereid

Distance from planet: 224 Neptune
  radii
Diameter: 340 km
Mass: 3.09×10^{19} kg
Orbital period around Neptune: 360
  days

PLUTO

Distance from Sun: 39.5 AU
Diameter: 2,377 km
Mass: 1.303×10^{22} kg
Orbital period around the Sun: 248
  years
Spin period: 6.39 days (retrograde)

Charon

Distance from planet: 16.5 Pluto
  radii
Diameter: 1,212 km
Mass: 1.55×10^{21} kg
Orbital period around Pluto: 6.39
  days

Nix

Distance from Pluto-Charon barycenter:
  41 Pluto radii
Diameter: 74 km
Mass: 4.5×10^{16} kg

Orbital period around Pluto-Charon:
  24.9 days

Hydra

Distance from Pluto-Charon
  barycenter: 54.5 Pluto radii
Diameter: 38 km
Mass: 4.8×10^{16} kg
Orbital period around Pluto-Charon:
  38 days

HAUMEA

Distance from Sun: 43 AU
Diameter: 1,436 km
Mass: 4.0×10^{21} kg
Orbital period around the Sun: 284
  years
Spin period: 3.9 hours

Namaka

Distance from planet: 48.2 Haumea
  radii
Diameter: 170 km
Mass: 1.8×10^{18} kg
Orbital period around Haumea: 34.7
  days

Hi'iaka

Distance from planet: 60.7 Haumea
  radii
Diameter: 310 km
Mass: 1.8×10^{19} kg
Orbital period around Haumea: 49.1
  days

INTRODUCTION

Time is the father of truth. Its mother is our mind.
—GIORDANO BRUNO

I WAS BORN IN NORWAY IN October, so half a year went by before I had my turn lying on my back in the soft grass, gazing into the sky after sunset. (Never disturb a baby who is staring at the sky.) Still, occasionally through the dark winter I would have found myself outside, bundled up in a pram on a walk from here to there. I have no real memory of it, of course, but I'm pretty sure that my first sight of the Moon was of a cold crescent set against dark indigo among a few sparkling gems—a vision that through my life has stopped me in my tracks. Since then, perhaps because of that, I've been a student of planets.

I have a more distinct memory of my daughter's first encounter with the Moon. She was born in the summer in a temperate climate. When she was ten days old, we carried her up the neighbor's hillside to enjoy the lunar opposition,[1] when the Moon was at its brightest; it had washed away all but a few stars and maybe a planet. The air was quiet and cool, and some insects were out. I shall never forget her awestruck little face in the dreamlight, peering from the folds of her cotton sling. She made a new sound like a word and reached her fingers to the pale white nipple in the sky.

From infancy we know the Moon, and we have stared at it and been moved by it, and awed by it. Astrologers say that its presence is carved into our personality, our spirit, and our soul. Millions of years of humans have evolved beneath its constant benevolent presence, giving rise over a million-year time scale to a collective human awareness in which the Moon is anchor of poems, stories, mythologies, astrologies, and religions.

Humans have understood the Moon in scientific and prescientific ways—the geometers, timekeepers, recorders of tides, and predictors of eclipses. Priests and oracles; architects and planners; farmers and hunters and fishermen. In pursuit of a scientific understanding of the Moon, we cannot hastily unravel all of that. Scientific arguments for its origin and evolution are awash in context. Far beyond any geophysical, astronomical, or cosmochemical analysis, the Moon has *meaning*.

To obtain a scientific understanding of the Moon, we must work our way back to the first academic studies of the natural order. That means moving back to a time when observations were tangible things such as diameters (a half-finger width) and positions in the sky, and when natural philosophy was an amalgam of ideas and manners of thought. Instead of modern pipelines of powerful analysis, science *way back then* was more of a general pressure of ideas expanding outward, a widening sphere of knowledge connected with the human-spiritual quest. As you read this book, keep in mind that you have the liberty to move on to a different paragraph or chapter as you please, aided by the illustrations that correspond to text in various chapters. Language is linear, but narrative need not be.

Science as we know it has always been there, although it has increased greatly in reach and been proportionately reduced in scope. Philosophers used to be the astrophysicists and atomic theorists, in days of yore. Astrologers were astronomers, the ones who applied and studied geometry, the measure of the world. Chemistry was alchemy, whose jars and beakers and athanors gave material and ethereal substance to astrology. The wheel of the Wu Xing, which cycles from wood to fire to earth to metal to water, and back again,[2] conveys a primal geology and chemistry: wood becomes earth by fire; metal brings water. Deities of ancient Benin, Mawu the Moon and her brother Lisa the Sun, procreated with every eclipse in astrophysical symmetry. Eclipses, comets, and other celestial events, as interpreted by Stone Age artists, are preserved in pictographs in the deserts of the world—systems of knowledge we can barely fathom.

Every system of thought blends the scientific and the sacred: how to best explain the natural world in the mind and in the heart. The explanations can't be *too* sacred, though. After all, the Moon has irregular markings, sometimes explained as a man or a rabbit, but not looking much like either. Is it a blemish or birthmark? Is it the goddess Selene riding sidesaddle, as some have said?

In the prescientific era, the imagination was able to run wild because no one had yet seen the Moon's surface with their own eyes, excellent though these might be. The air blurs things and we have only so many receptors. It was also noticed that the Sun has its own blemishes that come and go, *sunspots,* as recorded by Chinese natural philosophers who squinted through the smoke of forest fires; please do not do this.[3]

Beneath the fundamental cadences of planets—the day, the month, the year—were irregularities and intricacies that would take thousands of lifetimes, and the origin of astronomy, to figure out. To the animals on Earth, it doesn't matter that the cycles of the Sun and the Moon don't mesh,[4] that there are ten or eleven days left over between the twelfth full moon and the start of the new year. The day, month, and year set the fundamental beat, and on top of that are more complex cadences. But to humans who want to write things down and explain the specific order, the pattern matters.

The comings and goings of planets are significant and can be predicted. Mars is faint for more than a year and then grows red and bright during the *conjunction*, when it races alongside the Earth for a while, both planets on the same side of the Sun. It appears overhead, big and bright—a season of Ares that has often been foretold as a premonition of war. This would become a self-fulfilling example of augury, that Mars would signal troubled times. There was similar power in foretelling an eclipse, the legend of Thales of Miletus. On some nights, the stars would fall from the sky, burning up in streaks through the air. What did these portend? And how about great comets, their colorful tails blazing for nights on end for the world to see? Then, as now, there would be a competition to explain those things—my deity or yours, natural philosophy, magic, bullshit, and modern science.

Human culture goes back hundreds of thousands of years, and the first stories told may have been tales of comets more spectacular than any we have ever seen. Stories would have been told about a nearby star that exploded, that would have shone more brightly than the full moon for a week or two and then faded into a fairy ring lasting for decades. What was a stone-tool-wielding cave dweller to think? Every human being throughout the world would gaze upon it; nothing would ever be the same.

Although punctuated by strange and magnificent events, the movements of the Earth, Moon, and planets are generally harmonious. This established a romantic notion that what is *true* must be harmonious, or as young John Keats expressed it, "Beauty is truth; truth, beauty,—that is all / Ye know on earth, and all ye need to know."[5] The underlying harmony, the unfailing beating heart of the solar system, is reflected in our writing, painting, sculpture, music, and design, and in our science, which seeks a kind of regularity of structure.

The calendar is our attempt to capture the solar system's rhythms, the most fundamental being the *day*, which is defined as one rotation of the Earth, and for us humans happens to be one sleep cycle, which is as vital to us as food.[6] Each day in the English calendar has a planetary association: Sunday, Monday, Tiu's day (Mars), Odin's day (Mercury), Thor's day (Jupiter), Freya's day (Venus, Aphrodite), and Saturn's day.[7] Seven days a week times four weeks makes a *month*, which is approximately the orbital period of the Moon around the Earth.[8] Twelve and almost a half of those makes a year, which is the period of the Earth around the Sun. These rhythms are between the pulsebeat of the human heart, about a second, and the span of a human life, a thousand moons.

There used to be no need for clocks and calendars. *The corn will be ready in a fortnight.*[9] *I'll be back on the snow moon. It was the last summer when Mars was so bright.* You used the Moon and the Sun to tell time; nothing was ambiguous. Every bright star was familiar, and no newcomer to the night sky went unnoticed. The darkest skies you have ever seen—such was the night sky for everyone, from everywhere, when it was clear.

The lunar calendar is a living thing: when you try to write it down, it resists. After the twelfth full moon there are eleven days left over, more or less. After 365 days, there is a quarter day left over, but not quite, giving rise to the leap year and other complexities. What you do with these extra days and hours, and how you structure the whole thing, became the job of priests, whose first temples doubled as observatories, aligned to the Earth's orbit and rotation, east, west, and solstice. Somebody would be expected to come up with a divine order and provide satisfactory explanations for the variations of the year, the irregular markings of the Moon, and the meanings of comets and meteor showers. And none of these religions arose without prior context, the accumulation of human memory since the beginning of it all, awakened by some rare incomprehensible spectacle of the heavens.

Planetary scientists trade in stories, some of them true and others "to the best of our knowledge." Others we are trying on for size: bar-napkin estimations and a collection of what-ifs bounded by physics, geology, chemistry, and mathematics, yet made limitless by the fact that the only way to prove something is false is for somebody to champion that it's true. So the scientist's job is fact finder and provocateur.[10] Our planet was created in giant impacts—this is a fact—and our Moon was a consequence of that. Deduction from that fact produces ideas and images that border on the fantastic: a Moon that is ten times closer than it is today, ten times larger, a hundred times brighter in the sky,[11] its mottled, volcanic, heavily cratered face gazing down on the spinning Earth. The Moon would raise tides in Earth's oceans kilometers high, washing over the first continents, something we have not seen, that we deduce. Geology began. *Let the dry land appear: and it was so.*[12]

Now imagine two moons spaced overhead like your arms outstretched, a big one the size of your palm and a small one the size of your fist, orbiting among a ring of other clumps and smaller bodies. One would rise and the other would rise, like a mother and her cub, above the horizon of the rotating Earth. Once upon a time, there were.

One for whom a pebble has value must be surrounded by
treasures wherever he goes.
—PÄR LAGERKVIST, *THE DWARF*

Some kids grow up thinking dinosaurs are cool, or fire trucks or flowers; for me it was logic, math, and planets. I was happiest to be in my mind and go on walks—my head in the clouds, as my mother would say. Yet I also had a passion to discover and understand things, and this required wandering outside my bubble, first by teaching (which is the only way to really understand things) and then by studying to become a scientist working on planet formation and missions of exploration, the things that are the topics of this book.

After college I taught earth sciences to a class of high school freshmen. Although I didn't have an education in geology, I was able to come prepared to teach because the subject is so interesting. It draws you in, and soon you start looking around you with a different set of eyes. Our textbook was good reading, with excellent scientific illustrations and diagrams,[13] and I kept a copy at home. I would pore over the topography and bathymetry maps inside the front and back covers, the way I had pored over the massive 1960s atlas that we had in our living room when I was a kid—the one that had a vertical bargraph of the supersonic X-15 going to the edge of space, that showed how the Mercury astronauts would soon go even higher and shoot off into orbit. It showed Venus as slightly blue and larger than the Earth—an artist's mistake or liberty; it is actually yellow and slightly smaller. It also featured an illustration of how the planets were born: when another star collided with the Sun five billion years ago and extracted a cigar-shaped plume (also blue) that beaded up, forming big yellowish planets in the middle and tiny brownish-purple planets on the ends.[14] I had learned a lot, so I had a lot to unlearn!

It's one thing to gain a knowledge of astronomy and the laws of motion; it's quite another to learn about the otherworldly *landscapes* you can walk on. Although our textbook was called *Earth Sciences,* it ended with a generous helping of extraterrestrial geology— bizarro geology, as my first thesis student would call it—including pictures taken by a new generation of spacecraft that had landed on Mars, the Moon, and Venus, and from the Voyager deep-space tours of the outer solar system. It was the stuff of Carl Sagan's *Cosmos.* Most stunning, to me, were the wide-angle panoramas of the surface of Venus, where the Russians had landed half a dozen spacecraft[15] on a planet whose atmosphere is massive enough to

The first image transmitted from another planet. Venera 9 landed on the
hellscape of Venus in 1975 and performed a series of measurements that
would be repeated a half dozen times by the Russian space program in the
1970s and 1980s.
Ted Stryk, data courtesy the Russian Academy of Sciences

crush the hull of a submarine (these spacecraft were high-pressure
vessels) and hot enough to melt lead. On another full-page spread
there was a spectacular view from the heavily instrumented Viking lander, looking out over the morning frost of Utopia Planitia.
I had arrived at Mars and there was no turning back.

Mind you, this was five years before the internet,[16] so you
couldn't just click on images of things. Most libraries had dated
materials, and the closest thing to the World Wide Web was a box
of microfiche slides containing an entire journal archive. A modern textbook had unique value. Another big hit were the folding
stereoscopes and spiral binders of image pairs that would allow us
to "fly over" Earth terrains. (Unfortunately we didn't have any planetary image pairs.) We also had an eight-inch Schmidt-Cassegrain
telescope, some single-lens reflex (SLR) cameras, and several
laboratory-grade microscopes from university surplus. A friend of
the school had donated a black-and-white darkroom that we had
set up in the small lab between the classrooms. We had collections
of minerals to examine and scratch, and hand lenses for each student. The kids drew sketches and wrote in workbooks. We bought
a rock-polishing kit, dropper bottles with acid for detecting carbon-

ates, a set of sieves, and—something of a novelty—a 3D-textured topographic map of southeastern Arizona, which eventually got worn down from everyone's fingers, including my own, tracing their way through the mountains. We had a desert to explore.

Teaching geology awoke another memory, this one from around the time that I was two, of my father exploring a dry stream bed in the hills east of Los Angeles, poking around and overturning some rocks. Our car was parked under some sycamores; I remember the dappled light. It was a family picnic or outing. He smiled and beckoned me to come see something, and I remember his tanned face, his eyes squinting in the sun, his simple slacks and cool dry shirt, his elegant movements. I walked as best I could across the unfamiliar terrain and got to where he was pointing. A large branch was snagged in the bed and had caught some big rocks, making something of a sculpture. I think he was pointing to a black widow spider among the sticks, tangled in the shadows, showing me not to touch it. Or maybe it was a lizard, one of his New World fascinations. But what I remember most are the rocks! I don't think I had never experienced such things—broken and eroded boulders larger than my hands, green and white and black and pale red. The boulders in the shade were cold, and the ones in the sun were warm. There were pockets among the big ones where sand and pebbles and leaves had collected.

It was my first geology field stop. That memory would come around again when the Huygens lander returned images from a boulder-lined stream bed on Titan. I've always been drawn to such places.

HALF OF MY education in geology was getting ready to teach so that I'd have something to say. The rest of it was through osmosis, the transfer of ideas when you hang around and interact with good people, like the biology teacher[17] who was my mentor. I grew to understand that everyone has their own teaching style, and to appreciate the privilege of interacting with young minds. It was through this osmosis that I became, for the first time, familiar with the *structure* of science and the importance of controversial hypotheses such as Gaia and evolutionary bottlenecks, and the fossil records of deep time: the Carboniferous, the Archean, the Cenozoic.

I also taught physics to juniors and seniors; we spent weeks doing strobe light photography, setting an air-hockey table at an angle to derive Newton's equations of motion.[18] We ventured into calculus, which is best learned along with the laws of motion because their application is most intuitive (your brain does some form of calculus whenever you catch a baseball, I imagine).[19] Students ran behind skateboards loaded with bricks, pulling them faster and faster by holding a rubber band stretched to a constant length, to derive Newton's law that acceleration (meters per second faster, per second) is constant when the force is constant. They tinkered with donated equipment. They set up laser retroreflector experiments and built a wind tunnel with smoke tracers made out of cigarettes (bad idea). We learned pinhole photography, with each student making his or her own camera; this would teach them about geometric optics and something about laboratory procedures, developing prints in the darkroom.

This was a cool, pitch-black room for developing negatives, with a dim red light and a projector for exposing prints, with its bank of filter wheels from yellow to purple, and a drawer full of dodging tools for customizing the brightness of an image. There were trays of developer that you would prepare to the right concentration and temperature. You would immerse your print for the specified number of seconds before washing in the fixer. Today everything is data. Instead of darkroom chemistry, or making pencil sketches, we stare at monitors and tweak pixels. There's a growing separation between you and what you're studying.

One late afternoon when I had become a professor at a university, my friend and I set up a telescope outside my office for my intro to planetary sciences class, and the students came by to have a look at the Moon and Venus for extra credit. A dozen students were taking turns when a PhD candidate from the astronomy department[20] passed by on her way to the bus stop. *Oh, can I have a look? Please do! Is that the Moon?* No, the Moon is over there (pointing at the bright crescent some distance to the left)—that's Venus! She marveled like Galileo over the fact that Venus has crescent horns, like the Moon's, but that its image was diffuse, and that it appeared so yellow. She exclaimed, *I've never looked through a telescope before!*

Direct sensing of the photons that come from the cloud tops of Venus, reflected from the Sun, establishes a direct connection to the planet. Yet there is a distinct advantage of theoretical models and using digital data and computers. By indirect but powerful means they allow us to sense things we could never hope to sense, and to process vast streams of data in myriad ways. Increasingly, a computer manages, reduces, and even interprets the data stream before we ever see it; such is the reality of modern big data. Computers correlate stereo anaglyphic pairs into 3D images, allowing us to experience and even fly through complex data landscapes. Computers also make vast astronomy and planetary exploration data sets available online, democratizing science for anyone with access to the internet. Type "Enceladus" into your browser and a marbled ice world appears on the screen. Click on a lunar science page and fly down to the Moon on the Apollo 17 descent. Delve into the NASA Planetary Data System archives and be the first to study certain craters on Mars.

True telepresence is not that far off, where instead of meandering with your fingers over a 3D topo map, you will come along in real time on a virtual field trip as your avatar strolls through a lunar lava tube hundreds of meters high and a kilometer wide, illuminated by thousands of glow-pods, to observe a new settlement under construction ahead of the first astronauts, being printed directly out of lunar soil. It can be as real as you want to make it.

BY THE MID-1980s, space shuttle launches held far less interest than the historic launches of the Apollo missions. Shuttles weren't going to the Moon; they were going a few hundred miles up, into low-Earth orbit, to launch satellite payloads, test equipment and procedures, and lay the groundwork for the International Space Station. It was all very cool, and the launches were impressive to watch, but it was becoming routine—indeed, NASA *wanted* it to be routine, "Going to Work in Space."[21] Still, at the school where I was teaching, all of us were paying close attention to the tenth launch of the *Challenger* because among its crew was the first teacher in space.[22] One in six Americans were watching on live TV that bright January morning. The rocket exploded and the crew perished, crashing into the sea like Icarus.

After the pall of shocked disbelief,[23] the *Challenger* accident put NASA's human spaceflight program on hold for several years.[24] The shuttle was the only rocket NASA had for launching massive science payloads into space, so science was put on hold as well. The Galileo mission was next up on the launch pad, a massive but delicate space bird designed to spend years orbiting Jupiter. It had been made to exacting tolerances at NASA's Jet Propulsion Laboratory with the most advanced technology,[25] engineered for a seven-year tour in deep space that would ultimately last fourteen years.[26]

Already subject to the long delays typical of a flagship mission, Galileo now had to bear the brunt of Earth's gravity for three more years in storage, including the vibrations of being ferried by truck from JPL to the launch site in Florida, then standing down, and being taken by truck back to JPL for storage, then a few years later, back to Florida. The radioactive power pack was still strong, but a key mechanism failed. When Galileo was finally launched, the umbrella-type high-gain antenna for beaming the data back to Earth got stuck; several of the ribs failed to open. The exploration would have to depend on a backup antenna capable of transmitting less than 0.1 percent of the data. (Through the invention of image compression, what we now know as the jpeg,[27] most of the mission goals could be completed once care was taken to transmit exactly what was needed.) Little could I guess that five years later I would be a newbie on that adventuresome mission.

Not long after the *Challenger* disaster, a local geology professor led us on a field trip to the desert west of town,[28] a beautiful place rich in stark contrasts and subtleties that I had often wandered on my own, although more in the manner of William Wordsworth than James Hutton.[29] We crammed into the yellow bus, my classes plus biology and chemistry and their teachers, for an early morning ride over the small pass. To our delight it had snowed almost an inch before sunrise, so the cacti wore white hats—a precious sight! We pulled into a dirt parking area and the kids piled out and scraped up some snowballs and goofed off, and then we hiked a half mile down a trail that followed the wash.[30] We came around a bend—for some reason, this too is frozen in my memory—and there was a big tilted plate of sandstone and mudstone, red and tan, with deep ripples measuring a few fingers wide—part of an ancient beach. It had been buried and exhumed, the professor was telling us, and was millions of years from home.

I was absorbed by the texture of the rock. The words I heard,

Image of the surface of Saturn's moon Titan, taken on January 14, 2005, by the ESA/NASA Huygens lander.
ESA/NASA/JPL/University of Arizona

on this and other field trips, were beginning to clear away a kind of fog, an abstract stasis. An ocean margin used to be where we were standing, he was saying, a hundred million years ago. The dust and silt that was laid down as mud to form these ripple-beds came from a hundred miles to the east, eroding off the mountains that uplifted. Sediments were transported through long-gone valleys by ancient rivers, and as dust blown on millions of windstorms.

That's what I remember. I'm sure I got the details wrong, but it made *sense* . . . rivers running and eroding, oceans lapping at the sand—mountains rising . . . I did not understand the next part, that the ripples in the sand and mud would be buried by more mud, become part of an ancient seafloor, to solidify under more and more sediments, which would become rock, to be exhumed millions of years later when the mountains underneath all of *that* rose up. It was dizzying to think about. Space and time expanded.

The rays of the Sun beat down. After exploring a bit more, we took turns taking pictures, goofing off, pretending we were surfing on the beach, but I had started to feel a disquieting sensation, one that would grow over the next few days to become a revelation. I had walked here before, looked around at the hills and arroyos and across at the mountains, but had never *known* what was around me and under my feet. The last to head out, I slid my fingers again over the ripples from ten billion mornings ago. Reality was bigger than I imagined.

ACADEMICS WALK AROUND with one or two big questions in their heads, which is why they lose track of time and bump into branches. My own big question is this: If planets were born out of clouds of primeval material orbiting the Sun, then why aren't they more or less the same, like so many raindrops condensing from a cloud, or like so many bales of hay piled in a mowed field? Our two biggest planets, Jupiter and Saturn, are indeed somewhat alike— two spheres of mostly hydrogen (H) and helium (He). The next biggest, Neptune and Uranus, seem even more alike, giant spheres of mostly H_2O and H and He, although to be fair, we've never sent a dedicated mission to either one. These are giant atmospheric bodies. At medium sizes—what we might call human sizes, bodies we can walk on, at least in principle—the assortment of planets is as diverse as all the states of Europe, especially if you include bodies like Pluto and Titan that meet all the geological definitions of a planet.

Our planet, the Earth, started with a swarm of icy and rocky bodies around the Sun that grew into planets. The first-formed planets collided with each other, forming bigger planets and their moons, and producing debris that mixed with leftovers of the original swarm, a jumbled-up mess we know as comets and asteroids. After about a hundred million years, most of that excitement quieted down; the collisional kinks were worked out until planets no longer crossed paths. Any giant impacts that were going to happen had happened, and the system became stable as clockwork—almost.

This is a book about the origin of planetary diversity. Without getting ahead of my story, let's just say that nearly every planet and moon that ever existed in the solar system was consumed by something bigger than itself, and that makes all the difference in the world. Most planets are now inside of a gas giant (Jupiter or Saturn), or inside the Sun; others are inside of Uranus and Neptune. Two or three additional Neptune-mass giants are believed to have existed that were consumed by the Sun or else ejected to roam the galaxy. Diversity is a matter of perspective, of what's left: we don't behold any *ordinary* planets. Almost every planet that ever existed was consumed by something greater; what's left are the fortunate and the unusual survivors.

Human curiosity, guided by science and amplified by giant telescopes, has revealed hundreds of billions of galaxies, each with hundreds of billions of stars. There are more stars in the Universe than there are grains of sand on Earth—100 billion trillion, or 10^{23}—and we believe that most of them have planets. It's so special to be here on this intricate blue world that the questions rise up like balloons, almost unfettered by human reason: *What is reality? What is time? Are we unique?* The manifestations of geology to be found throughout the cosmos could make Venus, Enceladus, Io, and Haumea seem ordinary; we have only an inkling of whatever weirdness could be out there.

A SNAIL IS a geologist, feeling the roughness of a rock, its temperature and dryness. So is the raccoon, exploring the shallow bank where snails might thrive. Primate geology is more advanced: Does this rock scratch that one? Does it crack? Does it crumble or cleave more easily in one direction? What are its color, texture, and weight? How does it smell? These are *perceptible facts* accessible to any sentient being, facts that are immediately useful.

Imperceptible facts are things you can't sense but can detect with advanced technologies. The classic examples are the telescope, which enlarges your perception, and the microscope, which shrinks it. Both use glass lenses to modify and extend the sense of sight, and (in a classical instrument) your eyes are sensing the actual photons that were reflected or emitted from the star or planet or moth wing.

Modern research microscopes occupy whole buildings and telescopes have mirrors that weigh ten tons.[31] And we have moved beyond optics and magnification to collect *remote sensing* data in all manner of ways and at all electromagnetic wavelengths. A spacecraft in orbit around some faraway planet can acquire data streams from laser interferometers, thermal cameras, X-ray fluorescence spectrometers, neutron detectors, ground-penetrating radar, and so forth. Although incapable of making scientific decisions, a modern spacecraft has access to many more kinds of perceptions than an astronaut, who is at best able to see things through a visor and feel things through an awkward pair of gloves, but whose mind carries the *vision*—of the sort that has nothing to do with the eyes—for exploration, and whose body provides the interaction.

The small end of our immediate perceptual scale is the limit of touch and eyesight, about a tenth of a millimeter, a fine hair or a fleck of coarse dust. We are replete with much finer specialized sensors down to the molecular level. At the big end is the span of the human body, a meter or two.[32] A more subtle but equally fundamental perceptual scale is about 6 centimeters, the average distance between human pupils. Behind these offset cameras,

Comet 67P/Churyumov-Gerasimenko, about 4 kilometers from end to end, is the first comet to be orbited by a spacecraft. This image was obtained 28 kilometers from the center of the nucleus; at that distance, the frame measures 4.6 × 4.3 kilometers.
ESA/Rosetta/NAVCAM (CC BY-SA IGO 3.0)

the retinas produce stereo pairs that stream into the left and right halves of the brain. It's been estimated we might use half our waking brainpower merging left and right images in our visual cortex to produce our three-dimensional reality.

Consequently, for humans, some of the highest priority space exploration data are pairs of images acquired under identical lighting conditions (usually taken at approximately the same time) that are offset by about 7 degrees in order to mimic the stereo view of an object held in your hands, once you put on red-blue glasses.[33] Using our biological processing power, we can behold Olympus Mons as though it was right before us. Dragging a mouse, we can rotate the crazy-shaped comet nucleus known as 67P/Churyumov-Gerasimenko (67P/C-G) and overlay other mission data sets like spectroscopy and temperature on top of it, creating a colorful virtual object that we can study from all kinds of perspectives, even walking around inside the object,[34] expanding our perception of what's real.

Deeper imperceptible facts are unearthed in basement laboratories around the world, where precision instruments are used to measure individual atoms in bits of rocks and meteorites and lunar samples. Mass spectrometers that fill entire rooms can determine precise elemental abundances in one-millionth of a grain of sand (*a World in a Grain of Sand . . . Eternity in an hour*[35]). From this data, researchers are able to tease out the conditions (e.g., composition, temperature, pressure, timing, presence of oxygen or hydrogen) under which an individual crystal grew, and the compositions of its atoms. From those insights, we can put together stories and rein in or discard other stories—for instance, about how planetesimals and planets formed. The best of these analytical laboratories are as expensive to build and operate as an astronomical observatory, except that instead of looking *out*, they peer *into* a grain of rock, making discoveries at nanometer scales, not much larger than the atoms themselves.

It seems like magic, but the abracadabra is mathematics, and takes deduction to its limits. In science you follow the math to where it leads you. And quite often—almost always—you find out it's the tail that wags the dog—the tiniest or most faraway measurement imaginable that topples long-standing theories and creates new ones. Just as in the detective stories, that *one more thing*. Making the tiniest scientific measurements requires incredible technological precision—for instance, to harness the probing energy of a nanometer ion beam or to capture light from the farthest reaches of the universe.[36]

IT'S EASY TO fall into the trap of taking the geology of the Earth for granted. Here we are, inhaling and exhaling nitrogen and oxygen, plus some argon and carbon dioxide and other gases, exchanging some of that O_2 for CO_2 in our most vital biological function, respiration. We are familiar with oxygen in a gaseous state; however, almost all of Earth's oxygen is actually in the *rocks*, which is where the oxygen in the atmosphere would disappear if it were not liberated constantly by plants photosynthesizing CO_2 and H_2O.[37]

The silicate earth—everything above the metallic core—is almost half oxygen by mass, found in minerals like olivine, $(Mg,Fe)_2SiO_4$, which is two magnesium or iron atoms, plus one silicon atom, and four oxygens. (In cosmochemistry, rocks are examples of *oxides*.) Thinking this way, we have to delve into the story of oxygen a little, but don't worry if you don't understand it—nobody really does.

Each of these atoms was created in fusion reactions in the cores of ancient stars, a topic we'll get to a few chapters in. An atom is defined by the number of protons in its nucleus; for instance, an oxygen atom consists of 8 protons orbited by 8 electrons (to be electrically neutral) plus a number of neutrons. Stable atoms of oxygen can have 8, 9, or 10 neutrons, *isotopes* called ^{16}O (by far the most common), ^{17}O, and ^{18}O according to their atomic mass (number of neutrons plus protons). Chemically they behave almost the same and are interchangeable, except that ^{17}O and ^{18}O are somewhat heavier and a little more sluggish when it comes to reactions.

Their relative isotopic abundance is imperceptible without relying upon a mass spectrometer, an instrument that weighs the proportions of the individual atoms in a mineral.[38] But because they have different masses, the isotopes of oxygen become sifted and sorted. A molecule of H_2O made with ^{18}O is slightly harder to evaporate than a molecule made with ^{16}O. In an ice age, for example, when evaporated ocean water moves onto land and is deposited as snow onto an expanding ice sheet, the resulting glaciers hoard the lighter oxygen, leaving behind an ocean enriched in heavy oxygen. When fine sediments and carbonates are laid down at the bottom of such a sea in the middle of an ice age, the resulting rocks are therefore enriched as well. By applying mass spectroscopy to a deep-sea drill core, a graduate student can read the ancient climate like a graph.

Today the ice sheets are melting, so our sedimentary record is turning toward lighter oxygen. The lighter water trapped in the ice is finally flowing back to the sea. A record of what happened, on land and sea and in the air, will be preserved in future rocks. That's one rationale for obtaining samples of the sedimentary rocks of early Mars—not just to find fossilized microbes if they exist, but because those rocks, if sampled carefully, may preserve intricate evidence for Mars's own past oceans and glaciations. Even if fossils of living forms are never found, the characteristics of life can be identified in the isotopic organic chemistry of rocks, the very sort of record we puzzle over for life four billion years ago on Earth.

Meteorites litter the Earth. Nearly all of them are fragments of near-Earth objects (NEOs), which in turn are mostly fragments of Main Belt asteroids and comets. But a surprising number of them came from the surface of Mars. The evidence is this: Meteorites are grouped by their oxygen and other isotopic ratios, corresponding to parent asteroids that formed in diverse regions around the Sun.[39] One of these oxygen groupings is curious because it consists mainly of basaltic rocks. Oxygen could be found in any kind of rock, so the grouping must come from a planetary body big enough to be dominated by volcanoes. Mars is covered in huge volcanoes, and is relatively nearby; so is the Moon.

The clincher was the precise geochemical measurement of *noble gases* in what are now called the Martian meteorites—elements like argon and xenon that have complete electron shells and thus do not react. They don't make molecules, so when a lava crystallizes on a planet with an atmosphere—not the moon—they are trapped and can be measured billions of years later by heating the sample, providing a measurement of the atmospheric composition of Mars at that time. The noble gas ratios in these meteorites coincided with the atmospheric gases measured by the Viking landers in the 1970s.

It turns out that Martian meteorites are relatively common. The ones in museum collections would fill a wheelbarrow, and that's just a modern sprinkling. Looking back in time, as geologists do, it is likely that billions of tons of Martian surface rocks were delivered to Earth during the time frame of the origin of life four billion years ago, blasted off by the bombardment of comets and asteroids. Mars is a favorable launching point for target Earth. So why go to Mars to get more samples when meteorites deliver them for free? It's because the rocks that you want to study are still on Mars.

Most of the Martian meteorites are strong igneous surface rocks like basalts. We do not have any sedimentary samples, which would be biologically the most significant. Sedimentary rocks are usually weak and would break apart when ejected into space. If not destroyed by the ejection, they would break apart by thermal cracking while they are en route to Earth;[40] and if not then, they would explode when they crash into Earth's atmosphere at 20 kilometers per second. Basalts are more likely to survive all of that; sediments, not so much.

Still, the transport of sedimentary rocks from Mars to Earth is possible, and generally just requires the ejection of a much more massive unit of rock, which might have required an impact by an asteroid tens of kilometers in diameter, such as hasn't happened in a billion years. Four billion years ago, however, during the origin of life, these impacts were happening all the time. We need to go to Mars and collect these sedimentary rocks, not only to discover life if it is there (in living or fossilized form) but also to complete our understanding of the concept of *panspermia*, how life might get from here to there and across solar systems and perhaps across galaxies.

Since it's much closer, the Moon sends even more free samples to the Earth, also in the form of meteorites—over three hundred kilograms collected so far. These are found on ice sheets and in desert sands, for the simple reason that if they land where there are lots of Earth rocks, you'll be unlikely to identify them. They also are igneous rocks and have evidence of shock damage from their existence on the Moon and the cratering event that ejected them. Many of them have cosmic ray damage and ions implanted by the solar wind, indicative of eons spent near the surface of an airless world.[41] With the hundreds of kilograms of Moon rocks collected by Apollo astronauts, the reliable identification of lunar meteorites is straightforward.

But the measurement of oxygen isotopes does not discriminate Moon rocks the way it does Mars rocks. Moon rocks are isotopically *indistinguishable* from Earth rocks, to parts-per-million level, in oxygen and other elements. In fact, the bulk Moon is a decent cosmochemical match to dry Earth mantle. This has pulled the rug out from under the giant impact theory of Moon formation, because in the standard model the Moon ends up being made mostly out of the impacting planet Theia, which should be distinct from Earth, like Mars is. While computer modelers and theorists continue to come up with new mechanisms and proposals, the Moon formation conflict as a whole has retreated to the battle lines of the 1800s, when it was hypothesized that the Moon was ripped out of the Pacific Ocean basin. Maybe that idea wasn't so bad after all? The confusion is humbling.

IF YOU'RE LOST, retrace your steps. Go back to the river, to the source. Hydrogen, with a single proton, is the most common atom in the Universe, and next is helium, with two protons. They formed in the Big Bang. Oxygen is next, increasing in abundance as the cores of early giant stars exploded. One of the most common compounds is therefore water, H_2O.

What a remarkable molecule! Even before there were planets, water played a dominant role in defining the architecture of solar system formation. The disk around the Sun that formed the planets was originally mostly gaseous, the *protoplanetary nebula*. The pressure in the nebula was low enough that water existed only as a vapor, except where it could crystallize as a solid. In the distant reaches beyond 2 to 3 AU, water could solidify as frost, and these became the seeds that agglomerated further to become the so-called *cometesimals*, distant ancestors to comets. Nearer the Sun, temperatures were higher and silicates were the dominant condensates, forming rocky planetesimals closer in. This notion of an "ice line" has been a popular framework for understanding why there are terrestrial planets closer to the Sun, made of rock, and ice giants and gas giants and ice dwarfs farther out, although the architectures of exoplanetary systems have made us rethink that.

Bodies of liquid water were unable to exist until planetesimals grew large enough to provide gravity that would hold on to atmospheres and surfaces for water to condense on—the first oceans, heated from above if close to the Sun, or else heated from inside by radioactive elements. The prebiotic kitchens of life were hard at work. Today we bask on a planet whose surface conditions allow H_2O to endlessly transmute between liquid, vapor, and solid—the hydrologic cycle, around the so-called *triple point* of water. At pressures and temperatures within this range, water can rain, snow, thaw, and evaporate repeatedly, affecting all kinds of physical and chemical processes and cycles that define geology and biogeochemistry.[42]

Water, the universal solvent, also dissolves molecules from fine-grained minerals in the rocks and breaks down and transports sediments, facilitating and enabling an almost limitless suite of chemical and physical processes beginning with the breakdown of solids and their transport by fluids. In and among these processes, molecules learned to break apart and recombine in self-replicating ways.

Liquid water is also common in the outer solar system, where heat comes from within and below the surfaces of icy satellites. Tides raised by a planet cyclically deform a satellite's shape, causing ice shells and rocks to rub together, creating heat; this supplements the radioactive heating from the rocky mantle in the largest of these moons. This allows Europa, the Moon-sized icy satellite of Jupiter, to maintain a subsurface liquid water ocean equal in volume to the oceans of the Earth, shielded from the extreme conditions of space beneath kilometers of insulating ice. Enceladus, a 500-kilometer moon of Saturn, has liquid water spewing out of geysers, dispersing into space and crystallizing into a bright plume that becomes a faint ring around the planet. Ganymede of Jupiter and Titan of Saturn, the largest two moons in the solar system, also have subterranean global oceans, based on their size and composition and on how much heat is expected to be generated internally. Subsurface oceans are so common that we can say with some certainty that there are billions of ice-covered oceans throughout the galaxy. Would none of them spark life?

In our solar system, only the Earth has surface conditions near the triple point of water.[43] Let's suppose that this is a requirement for life. (If we think outside the box, maybe life doesn't require water, but could be based around another solvent—for example, liquid methane; for now we will stick to H_2O.) This strict assumption would not rule out Europa; it would just require pockets of gas beneath the ice. A planet where creatures might evolve to advanced consciousness, sometimes called sapience,[44] would need to satisfy other factors as well. It's possible, for instance, that the emergence of sapience requires perceiving something that is immensely beyond oneself, like a starry sky, or the Moon and Sun, or distant mountains. This might rule out smog-shrouded planets, or a deep, dark ocean miles below the crust, although even in our darkest oceans and deepest caves, whales and bats use sonar to obtain a detailed non-visual sensing of distance and space. Maybe the crust and mantle of a viable planet have to be active enough to build up mountains and continents and create ocean basins and volcanic islands, to provide enough niches so that biology can become diverse, and one of the many species climbs the evolutionary ladder to the top. Certainly the planet's sun has to be stable for billions of years. And maybe there has to be a moon, and not just any moon, but one that orbits close enough to create a total solar eclipse that blows some minds and lights the spark of sapience. Maybe a supernova has to go off nearby. Make the conditions as specific as you require, and then some, and ask, *How probable is that?*

FIFTY YEARS AGO we landed on the Moon "for all mankind." It was the first human presence on a planet beyond Earth. Since then, astronauts have been orbiting in laboratories in low-Earth orbit, while robots do the deep-space exploration. That makes sense, because robots don't breathe, don't need to return home, and have become expert at gathering images and other data from distant places, increasingly with minimal human supervision. Human and robotic, the first wave of solar system exploration has concluded, having surveyed all the archetypal objects: the Moon, planets, sat-

ellites, moonlets, rings, dwarf planets, ice worlds, comets, asteroids, and centaurs. The nitrogen plains of Pluto, the canyons of
Mars, the geysers of Enceladus, and the caramel swirls of Jupiter.
Massive telescopes are sent into orbit, and spacecraft are propelled
on repurposed war rockets to journey to Pluto and beyond, crossing the heliopause, where the solar wind is stalled by the pressure
of galactic space—the beginning of the deep end.

The next wave of exploration is just starting. It is a wave that
will not go much farther, but will rely on smaller and more capable spacecraft to visit hundreds of the strangest, most impressive
places and fill in the gaps, to complete our picture of the worlds
around the Sun—other moons, other asteroids, outer planets,
maybe Planet X. Hundreds of kilometers of lava tubes under the
lunar surface. The wave after *that*, I think, will be human again,
colonizing the Moon and Mars and perhaps Venus, and launching
artificially intelligent robotic emissaries to "nearby" solar systems.

The Moon[45] has been imaged to a half-meter resolution over
substantial regions. Similarly Mars[46] has been imaged globally to
about 6 meters per pixel,[47] and a third of a meter in targeted regions. All NASA science data are in public archives,[48] and in fact
nearly all of these images are waiting for any human to look at
them. Click on any of them and zoom in, and you're likely to be
the first person to notice something particular—a boulder that was
broken in half by a micrometeorite, for instance. Maybe you'll find
a new cave, of great interest to explorers and future colonists.

As for astronomy, new telescopes are coming online with capa-
bilities that were only dreamed of ten years ago. The Large Synoptic Survey Telescope (LSST) will produce 20 terabytes of data
every night, in a landmark search for short-lived events like moving
asteroids, transiting planets, and exploding stars.[49] This is causing a fundamental epistemological transition, in which there is so
much *big data* that we require artificial intelligence to process it

and even to make initial sense of it. We are teaching computers to teach us what the data mean.

This is what returns me to geology, where all you need is curiosity and opportunity and a guide. Pick up rocks and really look at them. Crack them open and use a hand lens to study them. Take pictures and read site descriptions. Take field classes where you measure dips and slopes. Work at landscape restoration. Look at the polished walls of an old stone building. Read guidebooks or bring along an app, to travel back in time on hikes and outings. Deep in the Grand Canyon, explore the Vishnu Schist, a metamorphosed basement rock whose name evokes creation itself, from humble beginnings as Proterozoic mud.

In all our searching and exploring, we haven't found anything we would call a second Earth. There are some candidates, but it's going to be decades before we know. What then? If an Earthlike planet is confirmed around one of the nearby stars—say, within 30 light-years away—would humans try to go? On a time scale of thousands of years, it seems inevitable that we should do this, that one of our great-great-grandchildren[50] would embark on a one-way multigenerational journey lasting centuries, off into the void. That sounds brutally dangerous and lonely to me, the stuff of hard science fiction. Long before then, I think we shall invent a second Earth next door, developing the ability to bio-precipitate the greenhouse gas CO_2 out of Venus's atmosphere, creating an exotic dreamscape that we can seed with life from home. We can ponder the ethics of that, altering the geology of planets, but the debate has become moot considering how we are willfully altering the biosphere of the planet that's our own.

IF EXPLORATION IS the yang of science, understanding is the yin. One can't advance without the other. We feel the thrill of explora-

Sunset over Pluto's mountains. Fifteen minutes after New Horizons made its closest approach on July 14, 2015, it turned around for this good-bye shot. The smooth expanse to the right is called Sputnik Planum, and to its left are Tenzing Montes, rising up to 3,500 meters. Hillary Montes are on the skyline. The backlighting highlights the haze layers in Pluto's tenuous but distended atmosphere. The whole scene is almost 400 kilometers across.

NASA/Johns Hopkins University Applied Physics Laboratory/Southwest Research Institute

tion in our bones, but there is a comparable satisfaction in understanding why it is so, how it came to be, and how it works. After five years of building instruments and a uniquely fast and light spacecraft, and making all manner of detailed preparations for its launch, and after ten more years and three billion miles of spaceflight, the men and women who worked on New Horizons[51] got to be the first humans to see the geology of the ninth planet, something that was a mere dot on a photographic plate from its discovery in 1930. That dot was resolved by the Hubble Space Telescope in the 1990s to reveal a pair of dots—a binary planet, a couple of blobs of light.

The spacecraft flyby in 2015 was done in fifteen minutes, an event that transformed Pluto and Charon from a few suggestive pixels into magical fairylands full of candy-cane mountains and chocolate-frosted seas, mountains cracking and folding in a strangely active geology of a kind we have never seen before, nitrogen plains overturning, and Pluto's beautiful bright heart. This after decades of thinking, observing, digesting, figuring things out, tinkering, planning, proposing, lobbying, designing, testing, and building. And so it is that the next adventures of exploration are quietly beginning in laboratories and research centers around the world.

We have surveyed only a fraction of the bodies in the solar system. We've never had a dedicated mission to Neptune or Uranus, and we've never landed on Mercury or on Ganymede, the largest satellite we know. What's it like down there, where the geysers are erupting on Enceladus? The Moon and Mars have vast caverns that remain completely hidden. These are places we can go. Yet for all the places we've been, we have a familiarity with only this one, the planet that gives us life. *Why is the world this way?* is too big a question. *Why are these worlds so diverse?* is more to the point. To answer that, we have to go back to before there were planets at all.

CHAPTER 1

RUINED STRUCTURES

Heaven's net casts wide.
Its mesh is coarse, but nothing slips through.
—TAO TE CHING

THE STUDY OF PLANETS HAS given rise to hundreds of celebrated philosophers and forgotten sages throughout the world. As with the Big Bang, there was no center to the expansion, but a few great visionaries stand out like signposts on the road that got us to this place.[1] The most transformative epoch in astronomy was in early Greece, so we could start there, or earlier still, in India and China—as if we have to start at all. As if we weren't just beginning. But for the sake of narrative, and because it's familiar, let's go straight to the time of Shakespeare, when the Copernican revolution had spread throughout Europe, and Johannes Kepler was the studious defender of that most dangerous of theorems, *De revolutionibus orbium coelestium,*[2] that the Earth goes around the Sun.

Although he would design some of the most versatile telescopes in his day, Kepler's research was grounded in pre-telescopic observations[3] with an emphasis on *astrometry,* the precise geometry of planets in the sky. Geometry, the measure of the world. The retrograde motion of Mars, going westward through the stars, and then eastward (and brightening), and then westward again, was explained as being due to the relative movements of Earth and Mars, like one horseback rider looking at another against the backdrop of the hills. While both horses gallop in circles, it appears that one is sometimes falling back, sometimes racing ahead, and sometimes sweeping a fast arc as it passes on the inside bend. With the Sun at the center.

This relative motion is the physical geometry of *parallax*, the same type of geometry that was used by the early Greeks to measure the distance to the Sun. Aristarchus of Samos and others had already figured out in the third century BC that the planets orbited a central fire, that it was many times farther away than the Moon, and that the stars were many times farther still. But empires fall apart; Greek knowledge was all but forgotten, and rediscovered in the West in a way that led to modern physics.

Kepler was bothered by systematic errors in the positions of the planets. These errors would go away, he showed, if the planets orbit the Sun in *ellipses* rather than circles, and if they move faster when they are closer to the Sun. It was brave to propose such a bold new geometry of the heavens, and the church did not take kindly to it. Only nine years earlier, the Italian philosopher and sometime astronomer Giordano Bruno had been executed for maintaining, before the invention of the telescope, that "innumerable suns exist. Innumerable earths revolve around these suns in a manner similar to the way the seven planets revolve around our Sun. Living beings inhabit these worlds." With no more evidence to back him up than existed in the age of Thales of Miletus, Bruno stuck to his guns and was burned at the stake in 1600.[4]

Kepler's own mother had come close to being burned at the stake as a witch, so he was closely aware of the danger of radical

An engraving of the cosmic boundary by an unknown artist, likely from the 1600s, as reprinted in Camille Flammarion's 1888 book *L'atmosphère: météorologie populaire*. The character resembles Giordano Bruno, who was martyred in 1600.
Camille Flammarion, L'atmosphère: météorologie populaire (Paris, 1888), p. 163

ideas. His approach to doing science was more determined and less flamboyant than Bruno's (although later in his life he would write what many regard as the first science fiction story, *Somnium Astronomicum*,[5] which envisions human voyages to the Moon). Why did the errors matter? Kepler's equations for planetary motion, later known as Kepler's laws, would be put into a physical framework by Isaac Newton in the form of *gravity* and *momentum*, and physics was born.

Lest you think of Kepler as some sort of driven advocate of facts, his culture was *prescientific*. (Today we are becoming *non*scientific, which is a different and sadder thing.) In the early 1600s there was no coherent theory of physics and no real path forward with a quantitative law. Natural philosophy was revelatory: *This is what I see, and this is what I observe to be true.*

Kepler discovered a crystalline-cosmological relationship that he would never let go of, that the six planets orbit the Sun according to the distances of spheres circumscribed by the five Platonic solids: the tetrahedron, the cube, the octahedron, the dodecahedron, and the icosahedron,[6] published in a set of essays and discoveries called *Mysterium Cosmographicum, Forerunner of the Cosmological Essays, Which Contains the Secret of the Universe; on the Marvelous Proportion of the Celestial Spheres, and on the True and Particular Causes of the Number, Magnitude, and Periodic Motions of the Heavens; Established by Means of the Five Regular Geometric Solids.*

In Shakespeare's day, one could not imagine there being another planet[7] any more than one could imagine there being an additional day to the week, so Kepler lived out his days comfortable in his received wisdom. Long after he was gone, this would all be upended when the first new planet since antiquity, Uranus, was discovered. This would ruin the *Mysterium Cosmographicum* and other theories, and confuse us about the days of the week, but shored up Newton's general laws.

And here we are today. Never in his wildest dreams would Kepler have imagined a telescope orbiting the Earth in space, bearing his name, one that would discover thousands of new planets of all sizes and orbits, many of them potentially habitable and even Earthlike. Nor would he have dreamed that every one of them would follow the laws that he set forth in *Harmonices Mundi*. I wonder which idea Kepler had more faith in, the one that he was so wrong about, his beloved *Mysterium*, or the one that would become the basis for the natural laws and much of modern science?

ALTHOUGH SHE WOULD not have considered herself a planetary scientist, the French-Polish physicist Marie Curie discovered the atomic nature of radioactivity, bringing a reality that would ultimately topple the edifices of nineteenth-century geology and lead to revolutionary ideas about planets and stars. She showed that radioactive atoms break down in decay chains, forming stable daughter products. On a time scale of billions of years, two common isotopes of uranium, ^{235}U and ^{238}U, break down,[8] becoming isotopes of lead (^{206}Pb and ^{207}Pb). Because U is relatively common in rocks, its breakdown skews the lead-lead isotopic ratios inside of crystals, changing in time, so that the ages of rocks can be measured with surprising precision.

Prior to the discovery of radiogenic lead, scientists had been thinking of the Earth and the Sun as tens of millions of years old at most, based on a convergence of arguments that we'll get into. Maintaining a minority opinion, the Scottish geologist Charles Lyell, close friend of Charles Darwin, was ushering in the quantitative science of sedimentology and arguing for a much older Earth. It would take billions of years to lay down all the layered outcrops he and his followers had begun interpreting as uplifted ocean basins. The science of *geochronology* advanced, with debates raging over the age of the Earth and its surface evolution, and ideas of biological evolution spearheaded by Darwin's monumentally important *On the Origin of Species,* which argued for an immense span of time for life to evolve.

Understanding the atomic nature of radioactivity and the relationship between uranium and lead allowed scientists to reliably

ALMA image of the young star HL Tauri, about 450 light-years from Earth, and its protoplanetary disk. The star is estimated to be only 100,000 years old, yet planet formation seems to be well under way, as emerging giant planets sweep their orbits clear of dust and gas.
ALMA (ESO/NAOJ/NRAO); C. Brogan, B. Saxton (NRAO/AUI/NSF)

calculate the age of the Earth. By the 1930s it was known, based on simple graphs of lead abundance, that some Earth rocks are at least two billion years old. The emerging deep span of geologic time was being discovered just as astronomers were discovering the deep span of cosmic *space.*

In the 1920s the American astronomer Edwin Hubble went on to discover that we are one of a teeming multitude of galaxies distributed throughout the cosmos, the modern view. He also proved that galaxies are receding from us in all directions, and that those farther away are receding the fastest. He deduced from this that the Universe is expanding isotropically, like dots on the surface of a balloon that is inflating. Each dot thinks it's the middle of an expansion, but in reality none of the dots are special.

He estimated the age of the Universe by calculating the time it would have taken for all the galaxies to get to their current distances, starting from some theoretical critical point when time and space began (the balloon, uninflated). This age (although we don't understand what it means) also turned out to measure in the billions of years, just like the sedimentary layers, just like the uranium-lead clocks in the rocks, just like the biology. It had become very real, very fast, that creation goes back further than we ever imagined, and goes forward beyond the end of species.

ISAAC NEWTON GREW up in the mid-1600s, familiar with Kepler's laws. One of his key achievements was to generalize those laws into relationships involving mass and time and space. He invented the *inverse square law of gravity,* which states that two objects are attracted toward each other in proportion to their masses, divided by the square of how far they are apart. Since its conception, this law of gravity has always proven basically true, and it is so elegant that one gets the feeling it has been there all this time, imprinted into nature, just waiting to be discovered.[9] But the law of gravity is not *completely* true, so it was *not* there waiting, but must be a human creation fitting our preference for simple theories. Neither is Einstein's general theory of relativity the end; it is another human creation and a perturbation of Newton's theory, leading to more accurate predictions and a different reality beneath the equations, just as subsequent theories—themselves further perturbations and progressions—will be needed to explain new data and better knowledge. But discussion about physics progressed in mathematical terms using explicitly defined quantities of mass, time, and distance.

The greatest scientific advancement derives from paying attention to small inconsistencies. It had bothered Kepler that circular orbits didn't quite fit. By the late 1800s, it was similarly troubling that the planet Mercury had an orbit that was precessing too rapidly. Mercury's orbit is eccentric,[10] and the perihelion point (when it is closest to the Sun) changes location slightly every Mercury year. Its orbit *precesses*, tracing a pattern like a Spirograph. Precession is mostly due to the tugs from the other planets, but it did not add up; there had to be another significant perturbation not accounted for by Newton's law. Could there be a substantial "ether," a substance pervasive through space, dragging on Mercury? No; Mercury would spiral in. Could an unobserved planet Vulcan be influencing Mercury's orbit? No; Vulcan would have been discovered.

Along came the physicist Albert Einstein, whose 1916 theory of gravity known as general relativity predicted a precession of Mercury's orbit that accounted for the discrepancy. Astronomers were happy and went on to the next discrepancy. But for physicists it changed their world view forever, adding a new dimension to the Universe. General relativity does not invalidate Newton's law; it gives a geometrical basis for it: the warped curvature of spacetime. Gravity is not a force; it is the gradient of a *potential field*. For most of us the distinction doesn't matter—Newton's gravity holds true to high enough precision to govern the day-to-day movements of the planets and moons, and me in my hammock, and even rockets that reach outer space.

HUMANS AS GIFTED as Newton have been around in every generation since the Stone Age. Fossils show that the cranium doubled in size around a million years ago, and what that was all about, we may never know. We used this new brain to produce stone tools with increasing sophistication—flatter, made out of better materials, with more defined cutting edges and designs. Better tools allowed us to exploit the energy-dense foods needed to pay for these energy-intensive brains. The fabrication of each tool was a challenging puzzle, as was every hunt, every migration to new grounds. We learned the properties of rocks, and the patterns of the Moon and stars and what would become known as planets.

The Enlightenment was a period of global awareness when the smartest humans were able to be brilliant. As Newton wrote in a letter to Robert Hooke, we are "standing on the shoulders of giants." Science was made possible by the establishment of a connected culture, one that allowed it to advance and to follow every new and detailed observation from near and far—a deep well in Syene, a cliffside in Taihang, the Magellanic Clouds. Also, scientists of the Enlightenment were born into a world that was ready for a new world view, relatively free from the encumbrances of doctrine,[11] organized around a system for vetting, debating, and transmitting collaborative knowledge and supported by formal reasoning, especially math, that would allow them to weigh the planets and measure the charge of an electron.

Saturn's giant moon Titan was discovered in 1655 when Newton was twelve. The movements of Jupiter's Galilean satellites were known for half a century before that, so he grew up knowing the quantitative aspects of satellite orbits around planets, and planet orbits around the Sun. He observed that satellites also follow a Kepler-like law—the closer they orbit, the shorter their period—so long as you account for the fact that Jupiter and Saturn are less massive than the Sun, so the pull on their moons is weaker. Matter *causes* gravity—that is Newton's theory. And matter is *accelerated* by gravity, causing moons to orbit.

Applying his law of gravitation to the periods of planets and satellites, Newton calculated the relative masses of Jupiter, Saturn, the Earth, the Moon, and the Sun. Kepler's third law states that the period of an orbit is proportional to the orbital radius to the 3/2 power (more distant orbits take much longer) divided by the square root of the total mass.[12] Measure the period and the distance of the orbit and from that equation you know the planet's mass. (You can test this with the table at the start of this book.)

Then from the masses of the planets and moons and from their sizes, Newton determined their *densities,* thereby characterizing the materials they're made of. Because the Galilean satellites orbit Jupiter significantly faster than Titan orbits Saturn, for a given distance, he was able to estimate that Jupiter is 1.5 times as dense as Saturn and therefore is made of heavier stuff, or else more compressed. He also weighed the Earth and showed that it is 3.5 times as dense as Jupiter and is most likely made of rocks and metals. He then tried to weigh the Moon by estimating the force, exerted by the Moon, that is responsible for oceanic tides on Earth—although this calculation was too tricky to be done with success just yet. In a few brilliant years he had proved that the planets are strikingly diverse in their compositions—a mystery that we struggle to understand.

Geophysics, the other half of planetary science, lagged a century behind Kepler and Newton. For a long time we knew more about the workings of mostly empty space and the movements of the wandering stars than we did about the Earth beneath our feet. That's because the sky is always visible, while the bulk of the Earth is hidden. The deepest oceanic trench is only 0.2 percent of Earth's radius—not even a scratch on an apple. We barely know the composition and structure of what's inside.

One of the clues is that it gets consistently hotter as you go deeper into the Earth, about 25°C per kilometer. In the late 1800s,

William Thomson, aka Lord Kelvin, came to the conclusion that this *geothermal gradient* is due to heat flowing from the interior to the exterior, which is cooler. The value of the gradient (e.g., 25°C/ kilometer) corresponds to how long the Earth has been cooling, like putting a cooked turkey into the freezer.[13] After infinite time everything would be isothermal (at the same temperature), so there would be zero temperature gradient. Thus, it serves as a clock.

Assuming that *In the Beginning*[14] the Earth was a just-solidified sphere, Kelvin estimated that the Earth solidified 20 million to 400 million years ago. Then, independently, he showed that the Sun would also be about 20 million to 60 million years old, based on how much heat it's been putting out (luminosity) and how much energy is theoretically available. Everything fell into place.

Why was Kelvin's estimate so wrong? In both theories, he did not account for the importance of nuclear processes. During crust formation, uranium and thorium and other radioactive elements get concentrated in granites and other deep crustal rocks,[15] where they break down over time in spontaneous reactions producing heat. We know this because of the abundances of daughter elements that are produced in the crust by decay chains, like radon and lead. This provides an important heat source that Kelvin did not consider. As for the Sun, the actual mechanism of heat production—hydrogen fusion—also lasts billions of years. For another thing, heat flows inside the Earth not just by conduction but by convection (tectonics), so the basis for his calculation was off.

By another coincidence, Kelvin's estimates agreed really well with the theory of lunar tides, developed in the 1870s by George Darwin, son of Charles. But unlike Kelvin, Darwin was fundamentally right about the physics, just off in the calculation. In what was certainly the most far-out geological theory of his age, George Darwin—the apple not landing far from the tree—proposed that the Moon was flung out of the Earth's mantle by an original rapid rotation. That would be consistent with the knowledge acquired in his day, that the density of the Moon is about the same as the mantle of the Earth.

There are two major components to Darwin's theorem, which we'll return to because it's vital to everything we know about the Moon. First, he assumes the Earth was once spinning so fast that a sizable lump split off from the Earth and began somehow to orbit. Second, he shows that this lump, the Moon, would raise gigantic tides in the Earth, and those tidal bulges would tug at the Moon. If the Moon was already a few radii away from the Earth, then the tidal force would swing it farther out, twirled by the spinning Earth like a lasso into higher orbits. A Moon that was not already a few radii away would fall back down, according to the theory.

Ignoring this problem for the moment, Darwin reasoned that the radius of the lunar orbit tells you the Moon's age, if you can calculate the tidal forcing over time. He estimated that the amount of time it took for the Moon to be tugged to its present orbit, 60 Earth radii, was about 56 million years, although in his paper he took care to note that this was a guess based on poorly known parameters. Because he overestimated the tidal force, his age was too small by a factor of 10, so his happened to jibe with Kelvin's result.[16] This is an extraordinary example of three convergent scientific estimates, derived from completely independent assumptions, that all proved wrong.

Kelvin spent much of the remainder of his truly brilliant career besieged by newfangled thought. By the 1920s, through the analysis of lead, which is a decay product of radioactive uranium, several rocks were estimated to be billions of years old. In a remarkable piece of laboratory work in the 1950s, geochemist Clair Patterson of Caltech showed that the lead system in Earth rocks matches that of several primitive meteorites, implying that the age of the bulk Earth—if you ignore all the decorations of plate tectonics—is

The Moon.
NASA/GSFC/ASU

about the same. He obtained an age of the Earth of 4.55 billion years[17] to a precision of about one percent, a value that stands.

The oldest discovered rocks that are native to the Earth are from the Jack Hills in Western Australia. These contain *zircons* that have been reliably dated[18] as 4.4 billion years old, from the first geologic era, the Hadean. By comparison, the oldest materials in the oldest meteorites are 4.5672 billion years old, where the debate is about the final decimal point. Most meteorites have ages within a few million years of this value, known as t_0 (*t*-zero). Comparatively recent ages can be determined by the measurement of short-lived radionuclides that have an ongoing production and decay. For instance, carbon 14 is created in the upper atmosphere by cosmic irradiation and gets mixed into everything that uses carbon; it breaks down in about six thousand years. That's one of the shorter isotopic clocks.[19]

In between, the eons from the Archean to the early Proterozoic are as vague as ruined sketches in the sand—there are few chronometers that can serve as reliable clocks.[20] But just because their history is not recorded does not mean these billions of years were boring. To the contrary, everything was happening—continents were forming, plate tectonics was beginning, life was evolving, comets and asteroids were colliding. It's just that we can't yet place these events in time. Whatever happened here, billions of years ago, has been consumed by the conveyor-belt recycling of plate tectonics and subduction, with bits and pieces surviving here and there, a hopeless puzzle to put together. Since the Earth was being bombarded mercilessly during the Hadean, some of the oldest Earth material also made it to the Moon, where we can go look for it.

LIKE THE RUINED structures of an ancient capital, collapsed monumental ideas get picked over and reassembled into longer-lasting theories. Ideas are never thrown out but are repurposed and assimilated into something that works better. So it advances. The vast collection of logical elements and relationships and facts, like Kelvin's thermal models and Darwin's tidal models, are like columns and pillars and capstones that went into developing a theory; you reuse them. The theory that ties them together might come apart, but the underlying elements are true.

Some monumental ideas are neither right nor wrong, because we don't understand how to answer them. Dark matter is one such idea where we are missing the big picture. Another idea, closer to home, is the geometric spacing of the planets. This seems to have a basis in reality, but maybe not, and it might have a physical underpinning, or maybe not.

In Kepler's *Mysterium,* the planets were said to orbit on spheres that are concentric to the Platonic solids, the reason being only the mind of God. Another mathematical progression became popular in the 1700s that had more basis in physics; we call it Bode's law.[21] Start with 0, 3, 6, 12, 24, 48, 96, each number doubling the one before (the exception being 0). Add 4 to each term and divide by 10, and they obtained the distances to the known planets quite accurately. The planets from Mercury to Saturn were predicted to be 0.4, 0.7, 1.0, 1.6, 2.8, 5.2, and 10.0 astronomical units (AU) from the Sun. (An AU is the distance from the Earth's orbit to the Sun, 150 million kilometers.) The actual planets are at 0.4, 0.7, 1.0, 1.5, 5.2, and 9.5 AU.

If you are on the hunt for finding patterns, you will find them— that is what humans do. You will also tend to forgive omissions when seeking patterns. For instance, you might not have noticed just now that Bode's law has a missing planet at 2.8 AU. The *Mysterium* worked better in that sense, with one sphere for every planet. The discovery of Uranus, confirmed in 1781, the first new planet since antiquity, was as spectacular as it was transformative. It proved to be much larger than the Earth, and its orbit, estimated at 19.2 AU, was close to the predicted value of 19.6 AU. Bode's law was confirmed, but it was not yet proven. It still lacked a physical explanation, and it was problematic in that it never ended. Also, there was the missing planet at 2.8 AU. The hunt was on.

One of the more organized squadrons of astronomers called themselves der Himmels Polizei (the Celestial Police), led by the Hungarian Baron von Zach. They divided the sky into twenty-four sectors and searched dutifully; other equally ambitious groups campaigned as well. But the discovery went to Father Giuseppe Piazzi, a man who had not set out to discover any missing planet, but who was working diligently on a new star catalog he was producing at L'osservatorio astronomico di Palermo.

In his log from January 1, 1801, Piazzi reported the discovery of "something better than a comet," a wandering star, perhaps a planet. He intended to keep the discovery a secret until he could confirm it with follow-up observations, but word got out and he had to race to get ahead of his own discovery. By February he had managed to estimate the planet's orbital radius, approximately 2.8 AU! But he lost sight of his finding as it moved into the daytime sky, because the Earth goes faster around the Sun. He was left with only a handful of observations to go on and no reliable prediction where to look. Without astrophotography, there was no proof he had even discovered it, just his notes.

This lack of confirmation became an astronomical crisis—the missing planet, as predicted, now lost! Once it should have returned to the night sky, nobody could find it, and there were doubters. This presented a challenge to Carl Friedrich Gauss, the brilliant twentysomething who solved the problem in a few weeks by inventing the mathematical method of least squares for predicting future data from past observations. (It would not be the first or last time that astronomical necessity led to profound mathematical advancement; a more current example recounted above is how the crippled Galileo mission helped bring about the invention of the jpeg.)

The method of least squares underlies much of modern data analysis and even artificial intelligence. Suppose you have a mathematical model, in this case Kepler's laws, predicting where Ceres will be at some time in the future. If you know precisely where Ceres was in the past, then Kepler's laws will say precisely where it will be. But in reality you have only a few measurements of where it was, and those measurements have errors. What is your best educated guess? Gauss came up with a predicted orbit of Ceres by minimizing the squares of the errors (hence the name) between the prediction and the observational data. He showed the astronomers where to look, and Ceres, as the asteroid would soon be named, was refound.

After all this hoopla, the new planet turned out to be much smaller than the Moon.[22] Other objects were soon discovered at similar distances from the Sun, where Bode's law had predicted, but none of them were larger than Ceres. Gauss referred to them as "a couple of clods of dirt that we call planets." The Main Belt, as it became known, was a collection of small bodies between Mars and Jupiter that were eventually believed to be remnants of disrupted planet. (Variants on that basic idea may be correct, although as we shall see, the story also involves the migration of Jupiter and Saturn, the origin of Mars, and the loss of hundreds of original asteroids the size of Ceres.)

Then Neptune was discovered, another giant, this time orbiting at 30 AU—the wrong answer. Bode's prediction is 39 AU. Then Pluto was discovered orbiting at 40 AU (on average; its orbit has large eccentricity), when it should have been at 77 AU. But even while Bode's law was falling apart, it was recognized that the idea of planets spaced geometrically speaks to a physical process. Newton's inverse square law is a geometrical law, after all; maybe it somehow causes planets to be formed with some power of the distance from the Sun. If one planet forms at a distance x, maybe its formation will influence planet formation nearby, setting up a gap to the next planet. If each gap was twice as distant, then you'd get a Bode's type of law, x, 2x 4x, 8x . . . Instead of discarding the law, we might modify it or seek an explanation in deeper physics.

We now believe, for reasons I'll explain, that the planets have migrated substantially from wherever they started. Therefore Bode's law, if it applies, is not related to where planets formed but to where they ended up. Also, at some point the geometric progression must end; you have to have a last planet; Neptune could be forgiven for being off by a bit. Today the active frontier in Bode's law is to look for geometric spacings of orbits and gaps in other planetary systems.[23]

THE NIGHT SKY hasn't changed much in a hundred million years. The Moon was a percent bigger, a percent closer, and the month was a day shorter. Tycho Crater had just been formed, draping its rays of ejecta across the nearside, still prominent today. But as long as mammals have been around, the patterns in the sky have been the same—except for the occasional comet or asteroid, and cyclical changes in the axial tilt of Earth (the direction of north), and the coming and going of any supernovas, red giants, or nebulas in the stellar neighborhood. Wind the clock back further, the way George Darwin did, and the constellations become unrecognizable, and the Moon grows five times closer. Earlier still, the Moon was ten times closer, and before that, twenty times, and ultimately there was the day when the Earth and Moon were formed.

Further back, there were earlier giant impacts, those that formed the bodies that collided. Before that, there was the birth of the Sun, the condensation of its mother cloud that made a family of stars, the origin of the galaxy itself. Here, eventually, all astronomy books must go, so let's take a short detour to the beginning of time, when the Universe began to consume itself. Quarks and electrons united in the first few minutes to become the first atoms, and so began the ascendancy of matter into recognizable forms.

As the cosmic dawn unfolded over the next few million years, random uncertainty caused some regions to be more dense than others, and their gravity would pull against the energy of expansion at a "local" scale, creating trillions of initial galaxies like a foaming storm-raged sea. As the expansion continued, the galaxies blossomed and the Universe calmed down. One by one the galaxies merged, not unlike how planets merged in giant impacts, until today[24] there are about a hundred billion galaxies.[25]

One of the first things you learn in astrophysics is that gravity is unstable. How and when it is unstable determines the structure, distribution, and masses of galaxies, stars, planets, moons, and comets and asteroids. If there had been way too much gravity, equivalent to too much mass, the Universe would have fallen back into itself, a collapsing bubble returning to singularity. (Maybe this happens a lot, of all the multiverses that have given it a try.) If there was not enough gravity, then the early explosion might have expanded continuously, without clumping (maybe this also happens a lot, if you believe in multiverses, or maybe whether or not you do). Instead, the Universe (at least, *our* universe) was created with gravity in such balance that locally denser regions, but not the whole thing, fell back on themselves, in a cascade of sizes determined by the tension of creation.

Getting back to planet formation, imagine there is a theoretical infinite cloud of molecular hydrogen and helium, ready to start forming stars and planets. Its gravity makes it want to collapse, but temperature and pressure resist it. A small perturbation (one region slightly more dense than another) has more mass, so it has more gravity. This means that as the cloud cools it breaks up into blobs of a certain size that collapse further to become stars.[26] We believe that in such a process the Sun was born, part of a birth cluster of hundreds of stars that was sheared apart during the twenty orbits it has made around the galaxy, each orbit lasting a quarter of a billion years.[27] They have since been all mixed up like berries in the muffin batter, so that only a few of the nearby stars may be related.

The composition of the Universe was originally hydrogen and

helium and a trace of lithium, produced by the recombination of baryons following the Big Bang. Things started to become compositionally a lot more interesting deep inside the first stars, almost as though there was a game plan to get started right away, cooking the first batch of heavier elements like oxygen and silicon and magnesium that would be required for terrestrial planets and ultimately life. It really is uncanny.

The first stars were born big, and in their cores, heavier elements were gestated in the process of *nuclear fusion*. This is the same kind of reaction found in hydrogen bombs, but in a star it is caused by the constant application of unfathomable pressure and heat, reaching temperatures of tens of millions of degrees. Fusion is happening inside the Sun, converting six hundred million tons of hydrogen per second into helium. Four million tons every second *disappear,* mass converted to energy via Einstein's equivalence $E = mc^2$, where m is the mass and c is the speed of light. A sunlike star will sustain its nuclear fusion reactions for about ten billion years, according to reliable models, so we have about five billion years to go.

The more massive stars that formed in the first generation were not so lucky. They burned hundreds of times hotter and faster, and when their fuel ran out, their cores collapsed and they exploded, billions of them like so much popcorn going off, synthesizing fountains of stardust in the process—carbon, nitrogen, oxygen, silicon, magnesium, phosphorus, iron—from which rock and ice are made, and planets and oceans and people. The phenomenally intensive, cataclysmically radioactive phenomenon of a collapsing star is called a *supernova,* and it's in the expanding shell of this popcorn that the magic happens and the goodies of cosmochemistry get made. But for now we'll think about sunlike stars, which can make long-lived planetary systems.

Any collapsing protostellar blob has greater random motion one way than another. As it collapses, it therefore starts to spin, for the same underlying reason as Kepler's law, gathering *angular momentum* as it contracts, the total mass times the rotation. Matter near the center spins faster, causing the blob to flatten out into a protoplanetary disk, rich with ice and dust. The center condenses into a spinning protostar that will soon ignite under fusion.

The rest of the story is how and when the nebula goes away, and how this disk breaks up into planets under the influence of the newborn star. Imagine a river that has a small whirlpool trapping leaves and twigs and water spiders, out beyond the bank, a place in the flow where the damselflies relax. Planetary accretion begins like that, at the balancing point where angular momentum causes things to swing away, while gravity causes things to hold on.

Planet formation is a story of letting go: of angular momentum, of gas, of satellites. It is an intricate and magical machinery, a universe that clumps into a web of galaxies, whose gas and dust lanes clump further into billions of blobs, each blob containing the potential for one or two radiant stars that provide light and heat to a system of planets. We have been able to figure out the details for how that happens because we can use telescopes to observe the stars around us, many of them sunlike in mass and composition, at various stages in their evolution—like hopping on the subway and seeing babies and old people and shoppers and commuters, all walks of life.

IDEAS ABOUT ROTATING disks breaking up into vortices and subvortices originated when the best telescopes of the early 1700s were good enough to reveal the Milky Way as a band of discrete stars. These were either tiny compared to other stars, or many times farther away. Stars in nearby galaxies could not yet be resolved, but some astronomers speculated that several smudges in the sky were distant bands of stars. Europeans had never seen the Magellanic Clouds in the Southern Hemisphere;[28] these had a texture like the Milky Way. Andromeda[29] and a few other nebulas also appeared a bit "milky," and seemed to have a spiral structure.

The German philosopher Immanuel Kant developed the theory that these were *island universes* (what we now call galaxies), and posited that the planetary system around the Sun formed originally as one of these spirals, with planets forming as condensations in a *protoplanetary disk.*[30] Although the details have proven complicated, Kant was right about the idea that angular momentum will flatten out a collapsing gas cloud into a protoplanetary disk. But the idea that Andromeda and other nebulas are planetary systems proved too simple. I don't think he could have suspected that each one is actually a hundred billion solar systems whose stars are glowing.

One immediate problem with the theory of nebular condensation was that the Sun would end up spinning on its axis every few hours, the way a figure skater goes into a spin. But nearby stars have rotation periods of one to ten days, and the Sun rotates only once every twenty-five days. Jupiter, a mere 0.1 percent the mass of the Sun, has twenty times more angular momentum in its orbit than the Sun. In fact, if you somehow reeled all the planets into the Sun, gathering angular momentum as you did so, you would change the mass of the Sun by only 0.2 percent, but you would spin it up to one revolution per day. How did the angular momentum leave the Sun?

The answer may lie in the intense magnetic fields of young stars, which rotate with the star the way the Earth's magnetic field rotates with the Earth. As the powerful magnetic field of a young star rotates, it sweeps through the dusty protoplanetary disk that has been ionized[31] (electrically charged) by stellar radiation. The star's powerful magnetic field interacts with the charged dust and plasma, grabbing on to it like a gigantic disk brake. This clutching causes turbulent heating as the field grabs charged material, speeding the inner disk up supersonically and transporting material away from the Sun, opening an inner gap, and causing all kinds of mixing and chemistry. Since every action has an equal and opposite reaction, the star's rotation slows down.

Another inconsistency in the original model led to the theory of *giant impacts,* not just to form the Moon, but to form planets at all.[32] By the early 1900s it was understood that you cannot make an Earth-mass planet directly from the protoplanetary disk, because the Sun's influence would prevent such a relatively small amount of matter from accreting. The same mathematical relations that indicate that a cooling molecular cloud will fragment into stars tell us that the protoplanetary disk could not have clumped into parcels as small as one Earth mass. The Sun's gravitational influence would disrupt an Earth-mass clump at 1 AU as fast as it tried to form.

Although it was abandoned for a while, replaced by the hypothesis of a stellar collision, the nebula model came back in full force in the post-Apollo era to include an increasingly complex series of mechanisms that are essential but poorly understood: the condensation of early planetesimals in the presence of the nebula, the dissipation of the gas, the merger of planetesimals into ever-larger embryos and oligarchs—the dominant precursors to planets—and lastly, the collisions of oligarchs in a late stage of giant impacts.

THE 1900S BROUGHT geographic and cultural shifts to the astronomic scene, and the same industrial-age scientific and economic advances that enabled voyages to the Moon. The American astronomer Edwin Hubble had primary access to the world's first hundred-inch telescope on Mount Wilson near Los Angeles and observed the brightest stars in these compact faraway smudges that would later be known as *galaxies.* He was able to use the brightness of the stars to estimate their distances, thanks to a 1908 paper by another American astronomer, Henrietta Leavitt.

Actually, only men could be astronomers. Leavitt was one of a team of female assistants at Harvard College Observatory; her tasks included long calculations and cataloging stars in photographic plates. Working on the Magellanic Clouds, she took interest in the bright Cepheid variable stars, which get dimmer and brighter repeatedly with time. (This is thought to be because the ionized helium layer at depth oscillates between being opaque and transparent, so it heats and cools.) She found that Cepheids whose brightness varies over periods of months are much brighter than those that vary every week. She plotted this up on a graph, and ever since then, all you have to do is measure a Cepheid variable's period and you know to a certain precision its intrinsic luminosity and therefore, how far away it is. She had discovered the first *standard candle*.

With Leavitt's research and other papers calibrating this and other standard candles, and with primary access to this new gigantic telescope, Hubble resolved many individual stars in these starry nebulas. Although he frequently mistook clusters of stars for single stars (making them appear much closer), he was able to show in 1924 that Andromeda and the rest are thousands of times more distant than the stars of the Milky Way. Leavitt passed away before the ultimate significance of her work would be revealed.

Not only were they more distant, but the more distant the galaxy, the redder its stars.[33] This inspired Hubble to propose that the Universe is expanding, everything moving away from everything else,[34] causing a "redshift" to the most distant bodies that are moving away the fastest. Space itself is expanding, so the waves of light get stretched out into longer, redder wavelengths. He computed the rate of increase of the expansion velocity with distance; the modern value of the Hubble constant is about 70 kilometers per second per million parsecs. (One parsec is 3.26 light-years, the distance light travels through space in one year, which is 9.5 trillion kilometers.) If the Universe is expanding uniformly, then the age of the Universe is simply 1 over that number. Once you convert seconds to years, you get about 14 billion years.

Back on Earth, the geologists welcomed the span of time that Hubble's hypothesis implied.[35] But it was disappointing that none of the visible nebulas were examples of solar systems that are forming. According to the standard candles, these nebulas were millions of times too far, and therefore millions of times too large, to be planetary systems. Impressive but abstract. If the nebula hypothesis was right, shouldn't we see some protoplanetary disks?

One possibility was that the process of planet formation had already played out everywhere, as in the Book of Genesis. Another possibility was that planet formation doesn't happen anywhere else but here. It turns out that neither of these possibilities is true. There are proplyds around nearby stars, but unlike galaxies, their existence is short-lived. And being made of ice and dust and cold gas, they are invisible except when illuminated by the star, and opaque when seen edge on. You don't just point a telescope at a million-year-old star and see its planet-forming disk; you have to deduce its presence.

Today we believe[36] that planet formation finishes around sunlike stars in a few million years, to a few hundred million years, in a process that is common but happens rapidly enough that you have to be lucky to see it happening nearby. It is a dark and murky epoch, interspersed by impact flashes as planets collide and grow, shrouded in opaque dust, emitting few visible clues to observers here on Earth.

Unlike the galactic disks that are brilliantly lit, protoplanetary disks are made of gas and ice and dust and do not emit light. But they are warmed by the central fire of their star, potentially to high temperature depending on distance and the opacity of dust and gas obscuring it. A protoplanetary disk (proplyd) is usually shaped somewhat like a doughnut, with the inner wall that faces the star getting to high temperatures, hundreds of degrees, emitting infra-red light (aka radiative heat). Although the star might look like any other twinkling dot of light, armed with an infrared spectrometer you can see what your eyes can't perceive, this warm glow from tens of light-years away.

A prism shows that sunlight is composed of colors. Although Newton thought there were seven discrete colors blended in various ratios (a defensible idea), there are actually infinitely many colors, a smooth transition known as a *continuum* that is associated with a certain temperature, in our star's case about 5,500°C. A *spectrum* is a graph, with blue on the left and red on the right, where you plot the relative intensity of each color. A star like the Sun plots as a bell curve centered on yellow, with lesser amounts of light at red and green and blue wavelengths. For stars that have protoplanetary disks, there is a second, smaller bell curve that peaks in the infra-red, that you can't see but a telescope detector can. This is a sep-arate continuum corresponding to something with a temperature of hundreds of degrees. So the spectrum reveals two sources: the star itself, which is at thousands of degrees, and something warm and extensive around it, an orbiting disk of dust that is heated by stellar radiation and energetic collisions and the decay of radioac-tive elements.

Infrared observations of stars are not so easy. The challenge is that we live inside Earth's atmospheric blanket and are awash in its thermal energy and we have to see through all that. While the main ingredients of the atmosphere, nitrogen and oxygen (N_2 and O_2), are mostly transparent, water (H_2O) is an effective absorber of infrared light, and so are carbon dioxide and methane (CO_2 and CH_4). Furthermore, the ground around the telescope is warm, as is the dome and the astronomer and the air, all glowing in the infrared (that's how night vision cameras work). The detector itself has to be cooled to liquid-nitrogen temperatures to render it dark enough in the infrared to see anything. Because of these complications, and the fact that reliable infrared astronomical sensors did not exist until relatively late in the last century, it was not until the 1980s that we had the first unambiguous detection of warm disks around stars.[37] Cautious optimism arose that we would soon discover exoplanetary systems.

At high-altitude observatories like NASA's Infrared Telescope Facility (IRTF) atop Mauna Kea, Hawai'i, observers peer through *atmospheric windows* in the continuum, infrared wavelengths where H_2O and CO_2 are somewhat transparent. Of course the best place to do thermal observations is in space, far above the warm, absorbing atmosphere, and away from massive radiating objects. This is the significance of the $10 billion James Webb Space Telescope (JWST), to be launched in 2021, an infrared telescope with a foldable 6.5-meter mirror that is sensitive from the visible out to 28.5 microns. To do its job, it has to orbit a million miles away from the glowing Earth.

For the same price as JWST you could build a visible telescope on the ground that is five times the diameter, and use modern adaptive optics technology to bring everything to focus. Resolution would be five times better, sensitivity a hundred times greater, and the facility would be easily serviced and the data transmitted by cables. But you still wouldn't see the planets that are forming, and that is because it is only in the infrared that we can see anything. Specific infrared absorptions also tell you about its chemistry, and thinking ahead, what a planet's air and land are made of. The same sensitivity to absorption by atmospheric H_2O and CO_2 and other molecules that make infrared astronomy so challenging makes it so valuable.

The original holy grail of the observations of young stars has been to directly image a planet-forming system. This has now been achieved for a few nearby planets: hot massive bodies in the midst of a clumpy disk. At ALMA, the Atacama Large Millimeter Array, dozens of individual 7- to 12-meter telescopes (the size of a backyard swimming pool) are spread out on mobile platforms over several kilometers of high desert in northern Chile.[38] ALMA images reveal distinct nested rings around stars, and gaps in the disks, and other structures that indicate the presence of massive planets orbiting the star, corralling the gas and dust into lanes. Direct imaging of an Earthlike planet is decades in the future, perhaps requiring a fleet of space telescopes coordinated in a widely spaced ALMA-like array.

Proof of the existence of exoplanets has been around since the 1990s, based mostly on two popular methods. First is the detection of the subtle gravitational influence that a massive planet has on its central star. A Jupiter-mass planet, orbiting much closer in than the planet Mercury—one of the very common "hot Jupiters"—tugs its star around in small circles every few weeks or months. When a star orbits around this "barycenter," it comes toward the observer for half of the year and moves away the other half (the planet's year, that is), and this produces a tiny periodic redshift/blueshift in its starlight, just like looking for the cosmic redshift, only much, much, much more subtle. So the wobble in the star causes a small Doppler shift in the spectral lines that only the most sensitive measurement techniques can notice.[39] This technique gives the *radial velocity* because what you are detecting is the star moving toward you and away from you, from which you subtract the orbital velocity of the Earth around the Sun at the time of the observation.

The other technique is simpler to describe: searching for *transits,* where a planet passes in front of a star, dimming its light. Transits were initially more difficult to prove as being exoplanets, since sunspots can appear to look a lot like transits, but they are now the premier data set in exoplanet discovery and characterization. Transit astronomy reached its apex with the Kepler mission, the space telescope that used a 95-megapixel camera to stare continuously at 150,000 stars for more than five years, discovering thousands of planets throughout the 2010s as each star's light dimmed and brightened as its planets occulted.

As a graduate student in the 1980s doing mostly theory, I recall the astronomers disappearing for long nights to the telescopes in the nearby mountains, their sleep schedules upside down. Yet despite all the excitement, there was a conservative skepticism, a hesitation against claiming that there would be planets for certain—you won't know until you do. (It could be somewhat similar today, where one hesitates to make claims that there is intelligent life elsewhere in the Universe, although a lot of scientists expect it.) There were all kinds of hints, especially the evidence for disks of gas and dust that agreed with theoretical models of planet formation. Everybody seemed to know that the moment was around the corner. Years went by. Finally, in 1995, a Jupiter-mass planet was announced orbiting with a period of four days around 51 Pegasi, a sunlike star about sixty light-years away.[40] It was the first confident radial velocity detection, a big wobble every four days. Within five years the observational baseline was long enough that dozens of exoplanets had been discovered by teams from around the world. The floodgates opened.

Today the number of planets is up to four thousand, so many that we can't follow each of them up with detailed characterization. There are not enough telescopes.[41] A few dozen of these new planets are in the *habitable zone,* the region where, depending on atmospheric composition, liquid water can be present at the surface, and life might flourish if other things also go right. And if we expand the habitable zone to consider tidally heated oceanic worlds orbiting

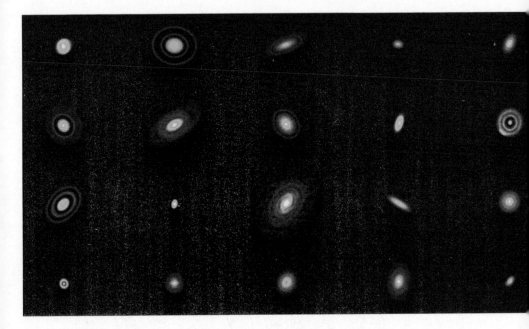

Twenty protoplanetary disks imaged by ALMA observatory.
ALMA (ESO/NAOJ/NRAO), S. Andrews et al.; NRAO/AUI/NSF, S. Dagnello

gas giants, the most famous example being Europa, then the habitable zone could extend as far as there are giant planets.

If all goes well, the JWST will characterize dozens of potentially habitable planets. It will not be able to resolve them, but will monitor them for evidence of satellites and seasons, ice masses growing in winter and shrinking in summer, changing cloud cover, and spectroscopic features indicative of the composition of the atmosphere—for instance, molecular oxygen O_2, which could be a signature of complex life.

We won't be able to clearly image any of these exoplanets, the way we clearly see the Moon and Mars and Saturn, until we send robotic spacecraft there a hundred years from now. But for a few dozen of our nearest neighbors we should soon know their spectroscopic variations from the visible deep into the infrared—data that will reveal the compositions of their atmospheres, their weather, and their big-picture surface geology: Are they oceanic? Do they have continents or ice sheets? It will be a tumultuous couple of decades following the JWST, and speculations will abound. Maybe by the 2050s a precisely coordinated fleet of space telescopes tens of kilometers across, acting as one giant telescope,[42] will enable us to image one or two second Earths, equivalent to the first sketches of Mars in the 1880s or Pluto in the 1980s; perhaps that would prompt the first robotic missions to go there. We have a long journey ahead, but the time scale for substantial advancement has been decades, not centuries.

TO MAKE PLANETS, with or without people, you need more than the right elements; you need the right proportions of elements to make the right molecules. So let's return to the big picture of the molecular cloud of hydrogen and helium and other gases and tiny grains of ice and dust that became the birth cluster of stars, including the Sun. Supernovas exploding in the neighborhood sent shock waves through space that triggered the collapse of the molecular cloud, and they also polluted it with stardust, giving us the stuff of planets. As it cooled and contracted, the cloud beaded up into hundreds of clumps. Gravity further pulled each clump together to become a star.[43]

Astronomers refer to any elements inside a star (or galaxy or molecular cloud, etc.) besides hydrogen and helium as the *metallicity*, which is a word for how many goodies it has available for making terrestrial planets.[44] Calling any element more massive than hydrogen and helium a *metal* comes from the fact that the ratio of iron to hydrogen can be readily detected in spectroscopy of the Sun

The compact, dusty molecular cloud Barnard 68 is half a light-year across and weighs about two solar masses. At the threshold of gravitational collapse, in a hundred thousand years it will turn into one or more stars.
Very Large Telescope at the European Southern Observatory (FORS/VLT/ESO)

and nearby stars. Stars appear blue, red, yellow, or in between, but if you put a spectrometer at the end of your telescope, which is a very fancy prism, you begin to identify forests of small gaps that are called *absorption lines*. Discovered by Cecilia Payne-Gaposchkin and other astronomers in the 1920s, these are fingerprints of the abundances of elements in a star, caused by those elements absorbing specific wavelengths from the continuum of photons trying to get out from the deeper layers of the star. If the continuum is like the notes from a trombone, then the absorptions block certain wavelengths, creating a timbre for each element.

The more distinct the absorption signature, the more concentrated that atom is in the photosphere. Assuming the Sun is well mixed, then its composition is 73.9 percent hydrogen, 24.7 percent helium, and 1.4 percent other elements, mostly oxygen (1 percent O) and carbon (0.3 percent C), by mass. Dozens of other constituents have been measured, and more than sixty detected. If you factor the mass per atom, this means that over 90 percent of the atoms in the Sun are hydrogen; similarly the C:O ratio is 0.55 in terms of numbers of atoms.[45]

Primitive meteorites have abundances of elements that are close to solar abundances. If you plot the elements found in the Sun against the abundances of each element measured by mass spectrometers in primitive stones like Allende and Orgueil, you end up with a more or less straight line. Except for the gases and other elements that didn't make it into meteorites, there is a one-to-one correspondence (i.e., same composition), with a handful of notable outliers. Each outlier and its isotopes tell a story about how the meteorites were made and about how stars go about making their planets.

If hydrogen and helium (H, He) are the builders of stars, then silicon, magnesium, iron, and oxygen (Si, Mg, Fe, O) are the builders of rocky planets, and carbon, hydrogen, oxygen, and nitrogen (C, H, O, N) plus a pinch of this and a little of that are the builders of habitable environments. So now our focus turns to carbon and oxygen, the second and third most abundant elements in the Universe. Both are basic products of the thermonuclear fusion reactions inside of stars, especially the carbon-nitrogen-oxygen (CNO) cycle. It may be that carbon is produced generally by all stars, while oxygen is produced more by massive stars that erupted early; if so, then the C:O ratio of the universe as a whole is increasing. But for now the stars around us typically have about half as many carbon atoms as oxygen atoms, and so does the Sun.

For this ratio, as the giant gas cloud cooled, H and H became H_2 (the most common molecule), C and O became CO (the most common *compound*), and then there was CO_2 and CH_4 and NH_3 and HCN and other "CHON" stuff, eventually condensing into ices. Once those reactions were complete, most of the carbon was used up and there was plenty of free oxygen left over. As was discussed in the preface, oxygen makes *oxides*, the stuff of terrestrial planets. One of these oxides is water, H_2O, the second most common compound in the Universe. Then came the minerals that make up the crust and mantle of terrestrial planets, like quartz (SiO_2), olivine (($Mg,Fe)_2SiO_4$), and so forth. We call these *silicates* even though their primary ingredient is oxygen, not silicon (Si), because—counterintuitively—the production of silicates is limited by the availability of oxygen. When the oxygen runs out, there can be no more rocks and no more water.

Oxygen is the key to all these terrestrial assets. So what happens around the small percentage of stars[46] where the C:O ratio is much greater? What a disaster! Now the carbon consumes *all* the oxygen in making CO and CO_2. With free carbon everywhere and no free oxygen left to make water and rock, what are their planets made of? The giant planets would still be H and He gas giants, but would be streaked with black graphite clouds, raining diamonds. The "rocky" planets would be more intensely weird. Instead of silicates, there would be carbides, carbonates, and solid carbon, and instead of water there would be hydrocarbons like methane and propane, CH_4 and C_2H_6. An Earth-sized *carbide planet* would have a large metallic core, and above that would be a mantle of silicon carbides instead of silicon oxides—for example, SiC instead of SiO_2. Above the mantle would be a crust of solid carbon in the form of graphite in its upper layers, and compressed to diamond around 10 kilometers down.

It's quite an image: if the crustal plates[47] were to buckle and fold, mountains of gleaming diamond would rise up! The graphite would weather off in the hydrocarbon rain, revealing stunning structures of clear crystal streaked with black, illuminated under a hazy light. You could live in a cavern, cheerfully lit, as spectacular as Superman's Fortress of Solitude as long as you sealed it off and filled it with breathable air. But outside your diamond hut, the planet would be bleak and poisonous. Hydrocarbons would rain from the smoggy sky, circulating in a hydrologic-atmospheric cycle much as water does on Earth, creating oceans and lakes and mighty rivers.

If active geology were to crack apart the adamantine mantle, the planet would become textured with deep hydrocarbon oceans, maybe rift valleys opening up where bizarre kinds of life could emerge, using methane and propane instead of water as the solvent. But if there was no mountain-building process such as plate tectonics, the result would be a completely oceanic planet where kilometers deep would be a soft graphite seafloor, a gentle dark muck where manta rays and tube worms could thrive among the bubbling seamounts. One can always dream.

Bizarre as it sounds, we might have an analogue to a carbon-planet environment on the surface of Saturn's smog-shrouded satellite, glorious Titan. A planet in its own right, ten times the mass of Pluto, Titan has hydrocarbon seas atop a basement crust of solid water ice forming substantial continents. But it isn't actually a carbide planet. Beneath the methane-ethane seas, there's plenty of water ice, and deeper still, a global water ocean that is thought to be heated by Saturn's tides, acting on the satellite's eccentric orbit.

On the oxygen-starved surface of Titan, however, the methane rain does fall, creating terrains that resemble some of the picturesque postglacial landscapes on Earth, dotted with hundreds of intricately textured lakes, several of them hundreds of kilometers long, with crenulated shorelines, islands, and bays.

TITAN IS ABOUT the same diameter as Jupiter's largest satellites, Ganymede and Callisto. Each is about 5,000 kilometers diameter, the size of Mercury. If Jupiter and Saturn are typical gas giants, with typical C:O ratios, then their largest moons reveal to us how the formation of satellite worlds might play out throughout the cosmos. Yet there's clearly something we don't understand. Although about the same size and bulk density, Titan differs in almost every geological aspect from Callisto and Ganymede. Callisto is a cold dead ball of ice and rock that never got warm enough to differentiate, while Titan is the closest analogue to an Earthlike hydrospheric system.

That's just the beginning of the geological weirdness surrounding Saturn. Orbiting closer in than Titan are five *middle-sized moons* (MSMs) between 300 and 1,400 kilometers diameter, some of them pure ice, others half rock with ice on the outside. Enceladus is one of the smallest but has geysers indicating an ammonia-rich water ocean down below. Beyond Titan's orbit are two more moons, Hyperion and Iapetus, both mostly ice. Hyperion looks like an eroded ball of pumice. Perhaps my favorite is Iapetus, half the size of the Moon on a distant inclined orbit, and made almost entirely of water ice. It is girdled with a 20-kilometer-high equatorial ridge. Half of Iapetus is blinding white, the rest of it pitch-black.

Their extraordinary geology and overall weirdness aside, what's most puzzling about these middle-sized moons is that Jupiter has

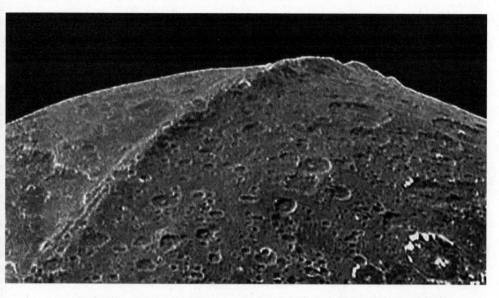

Geology tells a tale. A 20-kilometer-high ridge wraps all the way around the equator of Saturn's 1,500-kilometer diameter icy "walnut moon," Iapetus. Ideas for how it formed are rather crazy, but one of them is true, or none of them are crazy enough.
NASA/JPL

none of them. It has the four Galilean satellites, three of them in lockstep, and some scraps. Now there's a clue.

LIKE A COLD engine that backfires, the young Sun was sporadically hyperactive for the first few million years. Stars that are going through this phase are called *T-Tauri stars* after a well-studied active star in the constellation Taurus. When stars are finished with their birth pangs, they end up following a trend where the massive brightest stars end up huge and super hot, glowing blue, while the small stars end up cool and dim, glowing red. If you plot all the known stars on a graph with blue stars on the left and red stars on the right, and dim stars on the bottom and bright stars on the top, they cluster along a trend from upper left to lower right called the *main sequence,* with the yellow Sun right in the middle, and with lots of exceptions and branches where young stars are evolving onto the main sequence and old stars are evolving off.

The Sun, a very typical star, has been putting out light and heat at a pretty constant rate for 4.5 billion years. It's not so small, like a red dwarf that burns with a finicky light. Nor is it so large that it flames out in 10 million years the way that a blue giant does, becoming a supernova. Our star is a good star, and we've got plenty of fuel in the tank. It has ramped up gradually in luminosity by about 25 percent since the beginning, nudging up on the main sequence, but that's about it. Sure, we experience a coronal mass ejection every once in a while, when the Sun belches out a magnetoelectric bubble and washes the planets in streams of radiation.[48] But that activity is benign compared to what some other solar systems are regularly exposed to.

It won't be benign forever, though. In about 5 billion to 7 billion years will begin the Ragnarök, the final mayhem where the planets come unmoored. The Sun will become a red giant, expanding off the main sequence in a few million years, swallowing Mercury, Venus, and perhaps the Earth. Then the Sun will collapse, losing half its mass into space, an expanding shell of glowing gas that could be observed by nearby astronomers as a "nova" in their skies that would fade in a few thousand years. The Sun will lose its gravitational grip on the outer Oort Cloud, which will thereafter wander interstellar space like ghosts. What's left will collapse to become a *white dwarf,* a super-compact body that glows white from its gravitational energy—barely alive but highly luminous, the size of the Earth but a billion times more massive. We believe this is our solar system's fate, in part because the Sun is a common star with many examples at many stages of evolution to observe, and in part because theoretical understanding has advanced substantially and agrees with the observations.

After the red giant expansion is over and the Sun has become a white dwarf, planets and asteroids and other remnants of the inner solar system will proceed to spiral in, first by gas drag and then by tidal forces as the ultra-dense stellar remnant rips the planets apart one by one. In the end there will be a disk of terrestrial material, dominated by the disrupted mantles of Earth and Venus, that will spiral onto the surface of the ruined star. It's not just fantasy; astronomers see this in the spectroscopic signatures from several nearby "polluted white dwarfs," where rock-forming elements—magnesium, iron, silicon, oxygen—are present in the stellar atmospheres in abundances that correspond to silicate minerals like olivine—one last memory of terrestrial planets of yore.

PLANETS THAT FORM around stars much larger than the Sun do not have such an interesting existence. Massive stars burn at hundreds of millions of degrees, consuming H, He, C, N, O, and Si in ferocious internal fusion. The fusion reactions produce increasingly massive elements until the star reaches a critical state, exploding as a supernova, splattering its guts over a region several light-years across in a process that yields nearly all of the heavier elements. Any planetary system that may have formed is rendered moot.

Right now all eyes are on Betelgeuse, the prominent upper left shoulder of the constellation Orion. It's 600 light-years away, in the local neighborhood but thankfully not too close. It's eight times the mass of the Sun, about 10 million years old according to evolution models. Its explosion will compare to the Moon in brightness for a couple of weeks and then fade; if that sounds only modestly impressive, know that the intensity seen from 1 AU would be like watching a hydrogen bomb go off in your neighbor's yard. Over the course of geologic time, supernovas much closer

than Betelgeuse have exploded, irradiating the Earth and sometimes causing extinctions, but none of the nearby stars are about to
go off. The "kill zone" for this type of supernova is 25 to 50 light-
years, so Betelgeuse poses no threat.

Because it is relatively nearby and gigantic, Betelgeuse is the
first star to be resolved in telescopic images. Although the quality
of the images is poor, they reveal it to be a weird irregular spheroid, like a partially deflated balloon, that rotates once every thirty
years. It has a huge plume or deformation[49] that may be caused
by a global thermal instability. It sure seems ready to go off. But
the truth is, Betelgeuse must have exploded already in the time
of Kepler and Shakespeare if any of us are going to see the light
from that event.

When a massive star erupts, its thermonuclear bakery gets
busted open. The ashes of the fusion fire are distributed, and
He, C, N, O, Si, Mg, Fe, Ni, and other products of fusion expand
outward at hundreds of kilometers per second. These atomic nuclei, topping out at an atomic mass of 60, get saturated during the
expansion by a torrent of high-energy neutrons (the mass of protons without electrical charge) streaming from the collapsing core.
Some of the time a neutron colliding with a nucleus adds to the
mass; in a supernova expansion the result is the rapid synthesis
of more complex elements considered essential for life, plus many
that are radioactive, some with a *half-life* of only seconds, others
like ^{60}Fe and ^{26}Al breaking down energetically over a million years
while our protoplanetary nebula was forming, and others like ^{238}U
that are in it for the long haul, providing geologic heat for billions
of years.[50]

Here's what will happen when Betelgeuse explodes. Its core will
collapse in about a second to the radius of a neutron star, an object
so dense that a teaspoon of matter weighs a billion tons. It might
become a black hole. During its collapse it will spew out an es-

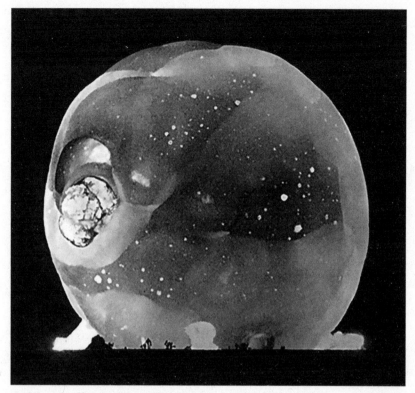

Only one million-trillion-trillionth the energy of a supernova eruption, the first-ever nuclear explosion in 1945 was captured 1/1000 of a second after detonation by Harold Edgerton's rapatronic camera with a shutter speed of 0.0000001 second. Joshua trees for scale.
MIT Museum, Edgerton Digital Collections

timated 10^{57} neutrinos, carrying away the energy so fast that the core collapses onto itself and a shock wave breaks out and tears the star apart. It will resemble the explosion of an atom bomb, only trillions of times more intense. Observed from Earth, Betelgeuse will brighten for days until it outshines its corner of the sky. In another couple of weeks it will diminish, then expand into a glowing nebula as it ejects a halo of gas irradiated by the compact beast at its center.

Supernovas pale in comparison to *kilonova* eruptions, which happen when two neutron stars get caught in each other's gravity web and spiral into a collision.[51] The two bodies are already so incredibly dense, each the mass of the Sun packed into the size of a 10-kilometer asteroid, that their merger pumps out gravity waves—ripples in the structure of space and time. Long predicted, gravity waves were detected for the first time in 2015 by a billion-dollar instrument known as LIGO.[52] Then, in 2017, a gravity wave coincided within 1.7 seconds with the arrival of a gamma ray burst detected by a completely different instrument—so a bang and a flash.

It is so remarkable that gravity waves and electromagnetic waves (e.g., photons) would travel for a billion years through space and time, seemingly independent (gravity and light being different), yet their arrival simultaneous. Perhaps the answer is trivial, or a foregone conclusion, but the synchronicity of gravity and light drew the Universe together for me in a way that matters deeply. It's as if there are ringing bells going off, these kilonovas from a billion years ago, a billion light-years away, and hearing them makes me feel connected to others who may be out there in the cosmos, in a way I didn't feel before—like seeing the Moon while thinking about someone you love, and reminded that they see it, too.

CHAPTER 2

ROCKS
IN A
STREAM

W*HAT DOES THE TIGER STAND on?* asks the skeptic. *It stands on the back of a giant elephant,* says the believer. *What does the elephant stand on?* asks the skeptic. *Why, it stands on the back of a giant turtle. Okay, what does the turtle stand on?* asks the skeptic. *Don't get smart with me, young man! It's turtles all the way down!* This familiar story, and the idea of an endless progression of turtles, is like the pursuit of infinity in Zeno's paradox: the arrow can never actually reach its target, because to get there, it has to first go half the way; and to get half the way, it has to go a quarter of the way, and so on ad infinitum.

Science arose out of the curiosity to get to the bottom of things—not just what the turtle stands on, but whether there *is* a turtle, even metaphorically, or whether we need to revisit the framing of the question entirely. Here's what Aristotle said about all this: "Since we must know the prior premises from which the demonstration is drawn, and since the regress must end in immediate truths, those truths must be indemonstrable."[1]

The regress must end in immediate truths. That sums up what can be called the Western belief that there must be a bottom to things, that there is an *indemonstrable basis*, self-evident, from which all facts are known or can be derived—principles of nature that simply *are*. In a top-bottom view, the role of science is to discover what the turtles stand on, but sometimes, instead of turtles all the way down, there might not even *be* a down.

THE IDEA OF Earth floating in space, unsupported by turtles or tigers, was recorded by Anaximander, a student of Thales who lived in the early sixth century BC. The more abstract or physical idea of a spherical Earth, with every point on its surface pointing *down*, which is the same as *in*, appears later that century with Pythagoras of Samos, a near-mythical giant of science and math.[2] Pythagoras traveled widely, and established a famous school that was visited by scholars from near and far. Cultures of deep scholarship had also emerged in ancient China, India, and Africa; Confucius (K'ung-tzu) also taught thousands of students, but with less emphasis on getting to the bottom of the turtles.

By 350 BC the spherical Earth had become a relatively widespread concept, and Aristotle wrote down several arguments that proved it. He noted that the Earth's shadow casts a circular arc on the Moon during a lunar eclipse; ergo Earth is a sphere or disk and the Moon is many Earth diameters away. He also noted that southern constellations are slightly higher in the sky, viewed from southern countries, a fact that is well explained if civilization spanned the surface of a globe.

How big is this sphere? The Greeks were prolific and kept good records; they also enjoyed expansive trade networks that brought them facts from far away, wonders of the world, and other remarkable things. One such wonder was a deep well in Syene, a city in southern Egypt where at midday on the summer solstice the Sun would be exactly overhead, so there would be no shadow at the bottom of the well.[3] On the same day at noon in Alexandria, on the Nile Delta of northern Egypt, the Sun was *not* overhead; a tall pillar in the city center cast a shadow at an angle of 7 degrees. Deep wells have to be vertical to within a fraction of a degree; otherwise they collapse, and the same is true with tall stone pillars. So what was going on?

Eratosthenes, a Greek philosopher who some say was the first geographer, was chief librarian of the fabled Library of Alexandria.[4] Upon hearing about the well, he reasoned that the pillar and the well must both be vertical, but according to a "straight down" that points to the center of the spherical Earth. From travel times recorded by runners, he estimated that the two cities were 5,000 stadia apart, where the ancient measure of one stadion is around 180 to 190 meters (it's the size of a stadium). The cities were 900 kilometers apart, and according to his reasoning, they were also 7 degrees apart on the arc of the round Earth. The circumference of a sphere wraps 360 degrees, so the circumference (π times the diameter) of the Earth is 360°/7° times the distance from Alexandria to Syene. (The degree, by the way, is a Babylonian invention; they fancied powers of 3, 20, and 60.) Solve for the diameter of the Earth, and the answer he obtained (in modern units) was 15,000 kilometers (which compares quite favorably[5] to the actual value of 12,700 kilometers).

In another brilliant application of the geometry of similar triangles, Aristarchus of Samos in the third century BC used the timing of the shadow-crossing of the lunar eclipse, an event that takes several hours, to calculate the diameter of the Moon. The diameter of the Earth was known. Therefore, assuming the Sun was much farther away, he knew the diameter of the shadow that the Moon has to cross during an eclipse. It takes four times longer for the Moon

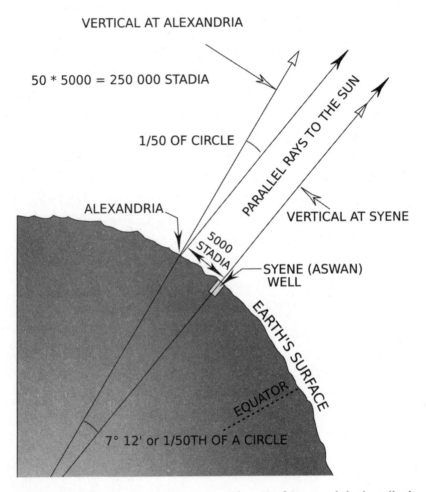

VERTICAL AT ALEXANDRIA

50 * 5000 = 250 000 STADIA

1/50 OF CIRCLE

PARALLEL RAYS TO THE SUN

ALEXANDRIA

5000 STADIA

VERTICAL AT SYENE

SYENE (ASWAN) WELL

EARTH'S SURFACE

EQUATOR

7° 12' or 1/50TH OF A CIRCLE

Schematic of the vertical well at noon in the city of Syene, while the tall pillar in Alexandria casts a shadow. Redrawn from Defense Mapping Agency Technical Report 80-003, "Geodesy for the Layman."
USAF, 1959

to cross through Earth's shadow than it does for the edge of Earth's shadow to cross the diameter of the Moon, so the Moon is 1/4 the diameter of Earth—about 3,700 kilometers, he estimated, nearly spot-on to the actual value.

The Moon spans half a degree in the sky in angular diameter.

It also moves half a degree per hour against the background stars as it orbits, so it crosses its own diameter every hour. From further simple measurements of angles, an object that spans half a degree has to be 110 of its own diameters away, whatever that diameter is.[6] The size of the Moon being known, the radius (distance) to the Moon was therefore known, 60 times the radius of Earth—another stunning achievement.

Why stop there? Greek and Chinese astronomers recognized that the Moon does not radiate light the way the Sun does, but that it reflects the Sun's light as it orbits the Earth. How far away is the Sun, then? It's obviously much farther than the moon, but beyond that, the measurements get difficult. Aristarchus realized that if the Sun is not infinitely far away, then the first quarter moon and the third quarter moon—the two moments of half illumination—would occur at angles a little bit shy of 180 degrees apart. And indeed, the Moon makes an exact D-shape a little bit early in the first quarter, and a bit later in the last quarter, but it is a barely noticeable discrepancy of much less than a degree. Aristarchus overestimated it as 3 degrees, perhaps believing it could not be much smaller; from this he estimated that the Sun is 20 times farther than the Moon. It's actually 400 times farther away.

Aristarchus also reasoned that stars, whatever they might be, are much, much farther than the Sun, because they do not show any parallax (apparent movement in the sky) when the Earth orbits the Sun. If any of the stars in the sky were only a few hundred times more distant than the Sun, then their apparent positions would change perceptibly during the seasons, the way that a vase on a nearby table appears to move relative to a distant wall when you blink your left and right eye. Since he could not observe parallax in any star, he determined that they must be thousands of times farther than the Sun. He was pushing it.

Lastly, since the Sun is twenty times farther than the Moon (he thought) yet the Moon covers it exactly in an eclipse, the Sun must be twenty times bigger than the Moon. He was right, but the measurement was off. The Sun is actually 400 times farther and therefore 400 times larger than the Moon, or 109 times the diameter of Earth.[7] This meant that the Sun, way up in the sky, is the center of the planetary system. The Moon orbits the Earth, which orbits the Sun along with Jupiter and the rest, which are surrounded by stars that are thousands of times farther still.

This was not conjecture; it was *real*, because geometry is real, and the data were simple enough for others to reproduce. The revelation that the Earth is a giant sphere, although small compared to the unfathomably gigantic and distant Sun, would have been decentralizing to a culture that at least formally maintained that the gods lived on Mount Olympus a few hundred miles to the north, and that the Sun was pulled on the chariot of Helios. To most people, the Earth was still flat, the sky was up, and an underworld of caves and subterranean rivers and lava flowed below.[8] The Sun and Moon would rise and set, the Moon would change its shape, and that was that. Like string theory and dark energy today, the new physical cosmology was not of immediate concern, to the extent that it was known, except perhaps for the times when a comet would visit their skies or a total solar eclipse would bring dire nightfall. Then they would pay attention.

ANOTHER MEDITERRANEAN LUMINARY was Archimedes of Syracuse, the third century BC mathematician, physicist, and engineer, who thought deeply about infinity. He is famous for the "screw of Archimedes," a corkscrew that spins inside a cylinder so that the rotation compels the water to go uphill. (Archimedes could have obtained the design from Egypt; it is widely used in agrarian societies.) True to form, he did not develop the screw for irrigation; he had to solve the pressing problem of bilge water that was plaguing his gigantic luxury warship. Due to its unprecedented mass and displacement, the *Syracusia* had begun taking on water as soon as it was put to sea.

Perhaps to settle a bet, Archimedes once set out to prove that the grains of sand in the world are not infinite. In an eight-page letter to King Gelon[9] that summarized his longer and long-lost treatise, he came up with an upper limit that it has to be less than. This frequently circulated letter became known as *The Sand Reckoner*. However large the Earth may be, it must fit inside the Universe. For the size of the Universe, Archimedes cited Aristarchus and concluded that the stars are about ten billion stadia away. Now he could put an upper limit on the grains of sand, but there was a problem: a number system had not yet been invented that could count that high!

The largest number in his day was the myriad, 10,000. One pail of sand already contains one myriad of myriads. So Archimedes had to come up with a new way of counting, and invented *exponential notation*, a concept also known to pre-Vedic scholars of ancient India. There is always a number that is arbitrarily bigger than the one before: 100, 1,000, 10,000 . . . Exponential notation allows us to probe the infinite, put limits on the countable, and characterize the infinitesimal.

The application of this concept requires high abstract reason. Creatures that are incapable of abstract reasoning perceive the Universe in a linear progression—1, 2, 3, 4, 5 . . . the way we count things and move our bodies through space and time. Two miles, three miles. Eight apples. Nine apples. The ticking of a clock. (Although we do count time in exponential terms—seconds, minutes, hours—we experience it linearly moving through it.)

Exponentiation gives rise to progressions that are distorted in space or time: each division has a constant *ratio* instead of a constant value, so it goes 1, 10, 100, 1000, in each case moving 0, 1, 2, 3 powers of ten (the number of zeros). Exponentiation is a transformative concept without which modern quantitative science would be impossible. You can span the smallest quantum distance (the Planck scale, 1.6×10^{-35} m) to the diameter of the Universe (about 100 billion light-years, or 10^{27} m) with just 62 powers of ten. Anyone can count to 62.

Armed with this new means of reckoning large numbers, Archimedes went about his calculation. As far as I can tell, he cubed his estimate for the diameter of the Universe to get the volume of the Universe, and divided that by the volume of a grain of sand[10] to get an absolute upper limit to the number of grains, 10^{63}. Note that he did not count the grains; he just showed that they were countable. He recognized that this did not mean you could count them, which is a very different thing. All the sand grains on the beaches of the world number only in the quintillions, of order 10^{18}. One year is 32 million seconds, so counting ten sand grains a second would take you ten billion years. The Earth and Sun would be gone. You would need billions of high-speed sand-counting machines to get the job done before you are dead. Sand grains, like stars in the sky, are countable only in principle, but that does not make them infinite. Is the distinction philosophical, or is it crucial?

In the way that a microscope is a backward telescope, Archimedes applied these ideas in reverse to consider the infinitesimal. He solved Zeno's paradox by clarifying that just because there are infinite terms in a sum does not mean you cannot add them. He proved that $1/2 + 1/4 + 1/8 + \ldots + 1/2^n + \ldots = 1$, which means that Achilles *does* catch up with the hare, that the arrow *does* hit the tree. His proofs were quick and enticingly geometrical. By cutting up a square into smaller squares, he proved[11] that $1/4 + 1/16 + 1/64 + \ldots + 1/4^n + \ldots = 1/3$. He came up with infinite series that led to the world's best estimates for π and the cube root of 3, estimates of great benefit to engineering and mapping and science. Not until the Enlightenment would deeper subtleties of infinity be discovered, leading to calculus, which is to modern physics what geometry was to the early Greeks.

A GEOMETRIC SERIES goes on forever in both directions: . . . 1/64, 1/32, 1/16, 1/8, 1/4, 1/2, 1, 2, 4, 8, 16, 32 . . . As you go to the left, you get to ever-tinier numbers, never quite 0, and as you go to the right, you get to ever-larger numbers, never quite ∞. This number line is called *base two*; the number of atoms in the Universe is about 2^{76}, so it would be 76 tick marks to the right. Because binary digits ("bits") can be represented as a simple on-off switch and still hold huge amounts of information, binary arithmetic became the basis for electronic computing.

Binary series are also used by geologists and bakers and builders and farmers, because it is natural to divide things into twice as big, half as big, and so on. There's 2×4 lumber and 4×8 plywood. A gallon is four quarts, a quart is two pints, a pint is two cups, a cup is eight ounces. A chain is four rods, a mile is eight furlongs. A cobble is a rock 64 to 256 mm in diameter, a pebble is 4 to 64 mm, and a boulder is 256 mm and larger. Sand is 4 mm to 1/16 mm, and dust is anything smaller than that.

If you sift through the sand at the beach, you come up with some gravels, a bunch of sandy nuggets, and a lot of fine sand—a geometric progression so far—but then not much dust. There is an overall constant ratio of sand to pebbles to cobbles—say, a hundred times as many of each size—so why does the progression end? Variations from the constant ratio tell a geological story. At the beach, particles of fine sand and dust are transported away by water, leaving coarse sand that we like to sink our toes into; the mud drifts away into the sea, to settle into the abyss.

The lunar surface has very little sand at all (although when you sieve it to get rid of anything *but* the sand-sized grains, these are very interesting indeed!). The upper 10 meters is mostly a fine powder, 20- to 70-micron igneous silicate dust (1/1,000,000 of a meter, a tenth the size of a grain of sand), plus a fraction of larger blocks and some gravels—a pervasive *regolith*. On the Moon the geometric progression continues until the material is predominately dust; unlike at the beach, dust production is happening much faster than it is being removed. This is because the Moon lacks rain or oceans, and as for a breeze, there's only the solar wind, which picks it off grain by grain. Since there is no atmosphere to filter out the small and common meteors, dust dominates the upper few meters of the surface.

Much of what we know about the lunar surface is hidden in the sizes and the textures of the grains. Because there is no atmosphere, small cosmic pebbles make it to the surface at full speed, creating pits the size of a shoebox; do this a trillion times and you have a softened texture. And while it is very unlikely that any individual astronaut will take a direct hit, this pelleting from space is something for long-term future colonists to be aware of.

Thermal cracking also breaks rocks apart on the surface of the Moon, in response to the severe day/night cycles. The Moon is a desert that gets the same amount of sunlight as the Earth. Without an atmosphere, and with day and night each lasting a fortnight, the surface temperature swings by 300°C in certain areas, roasting and freezing. Rocks expand when they get hot and contract when they get cold, causing thermal fatigue, which cracks down boulders. All the while, dust-sized meteorites come along and pockmark and break down everything to microscopic scales, softening the edges and creating the fine lunar powder. And constantly the solar wind, mostly hydrogen and helium nuclei escaping the Sun, gets implanted into the powder, changing its physics and chemistry

Image of the robotic Surveyor 3 spacecraft that landed in Oceanus Procellarum, photographed by Alan Bean 2 years later. Note the dimples from the footpad where the lander bounced, imprinting into a cohesive powdery material like cake flour.
NASA/LPI

and even leading to the notion of harvesting hydrogen for fusion power on the Moon.[12]

BEFORE THE TELESCOPE, and in the absence of any evidence to indicate otherwise, some philosophers imagined the Moon to be an Earthlike planet with oceans and even people. Its dark and mottled features, just barely distinctive to the naked eye, were called maria (from Latin *mare,* sea) and had oceanic names like Mare

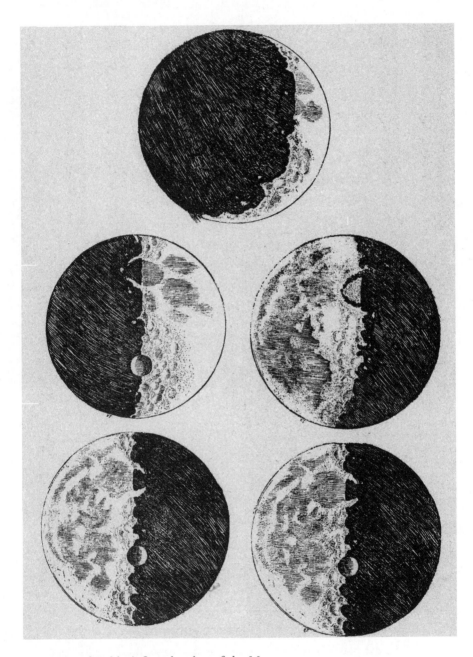

Some of Galileo's first sketches of the Moon.
Galileo, Sidereus Nuncius *(1610)*

Tranquillitatis, the Sea of Tranquility, and Oceanus Procellarum, the Ocean of Storms. By the time the telescope was invented, it had become clear that maria are actually plains, not oceans. This didn't imply that the plains were dead—some observers continued to attribute their dark color to vegetation.

Galileo introduced the geography of the Moon in 1610 with the publication of *Sidereus Nuncius* (the *Starry Messenger*). But it was the English philosopher Robert Hooke who first gave us lunar geology. Everyone should read his *Micrographia* (1665), or at least skim through it and read some passages out loud, because it is such a delightful romp after the invention of the microscope and telescope of all the things you can observe with these new marvels. It is captivating in a manner that transcends being a five-hundred-year-old work. He describes the manufacture and use of his state-of-the-art instruments; that his microscope requires a south-facing window and a large spherical jar filled with water to concentrate the light. The setup can be used only on the brightest days to get enough light on the stage to see things clearly. He proceeds to report with boyish delight that the tip of a sewing needle is actually quite blunt! He includes a detailed sketch, like a nine-year-old would for the science fair. He reports in detail, with other excellent drawings, on the anatomy of water bugs and fleas.[13] He explains why charcoal is black—it's not what you think.

Micrographia is mainly about Hooke's discoveries with the microscope—for instance, the "infinite variety of curiously figur'd Snow"—as well as some ruminations about physics ("the Elastick power of the Air"). But he concludes with a handful of pages describing his latest observations of the Moon, using "a thirty six foot Glass . . . the breadth of the Glass being about some three inches and a half"—that is, a very long focal length telescope with a primary lens about the size of a soda can. He describes the lunar highlands as "cragged, chalky, or rocky Mountains" rising high above the plains. He notices that these mountains are all around the Moon, obeying their own definition of "down." The Moon, he writes, appears to be covered in loose material that is attracted to its own center, regardless of where the Earth is. Without further ado, he drops the bombshell that "there is in the Moon a principle of gravitation, such as in the Earth"—the basic principle that matter attracts matter, the principle of gravitation.

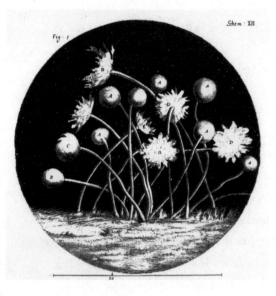

Microfungus *mucor*. Schematic XII from Robert Hooke's *Micrographia: or, some Physiological Descriptions of Minute Bodies made by Magnifying Glasses with Observations and Inquiries Thereupon.*
Robert Hooke, Micrographia (*London, 1665*)

Hooke's observational style is to keep a disciplined but open mind. On the Moon he views "a very spacious Vale" that "seems . . . to have its surface all covered over with some kinds of vegetable substances." That would have been nice. He also sees round holes, "some bigger, some less, some shallower, some deeper . . . incompassed with a round rising bank, as if the substance in the middle had been digg'd up, and thrown on either side." Craters.

A crater is a large bowl-shaped cavity in the ground that can form in a volcano, an impact, or an explosion. Massive craters are also dug out of the ground by mining, and all of these explanations have at various times been proposed for the origin of lunar craters. These ideas swirled around for a long time, but it was not until the Apollo era that this question was actually resolved. Hooke was able to describe them, but he wrestled with the question of their origin—a common situation in planetary exploration.[14] They couldn't be caused by impacting comets, because in his day, comets were thought to be planet-sized objects massive enough to obliterate the Earth "like a wasp nest into the fire," as Benjamin Franklin once described how that might turn out.

Asteroids were not conceived of until the first one was accidentally discovered in 1801. "It would be difficult to imagine whence such bodies should come," Hooke wrote when pondering the idea of lunar impacts. Still, he felt obliged to set up a series of experiments to test the hypothesis that something, perhaps long ago, had bombarded the Moon to make its craters. He and a friend or assistant fired musket balls into a thick bed of white tobacco-pipe clay, making craters from various angles and observing the result. He reported that their shapes and structures and rays compared favorably to lunar craters: shallow or hemispheric pits with raised rims and streamers of splattered ejecta. He was so close.

But the idea of bullets whizzing by in space proved too much, and Hooke concluded that the Moon's craters are probably volcanic, solidifying from a swirling melted state. His physics on this is most unclear, but he relates it to what he "observ'd in a pot of boyling Alabaster" (powdered gypsum) where vapors rising through the powder make "small pits, exactly shap'd like these of the Moon, and by holding a lighted Candle in a large dark Room." This is the observer's fallacy—if it looks like what you have seen before, that's what it is. Everyone is guilty. "In divers positions to this surface, you may exactly represent all the *Phænomena* of these pits in the Moon, according as they are more or less inlightned by the Sun."

The pages on lunar cratering in *Micrographia* may have been written after a visit to a friend's pipe factory. It is a thoroughly engaging piece of early science, and remarkably vivid and detailed. It is a small sliver of a much larger treatise that races through almost

Craters on craters on craters, almost a fractal. Apollo 11 image showing the 90-kilometer crater Daedalus on the lunar farside.
NASA

all of natural philosophy with these new eyes, the microscope and telescope. Still, had Hooke ended up concluding that impacts, not volcanoes, had created the craters on the Moon, it would have saved us a lot of time.

Imagine that for every 10-kilometer crater you can have 100 1-kilometer craters, and 10,000 100-meter craters, and so on, the number of craters increasing with one over the diameter squared (a *power law* with an index of 2). In this idealized case, a planetary surface can look remarkably similar at small and large scales, like a fractal. Imagine landing a spaceship on a fractal planet, monitoring your progress with a downward-pointed camera. As you come in for landing, soon all you see is a landscape of craters. As you get closer, your field of vision reveals smaller craters, but the larger craters go away because the camera footprint is smaller than they are. (Invariably you will be inside of some vast crater.) Statistically the number of craters in a given image stays the same,[15] so the image would keep looking kind of the same as you descended into the fractal. You would not be able to judge the distance as you landed.

In practice, of course, it's not that way. There are craters and blocks of preferred sizes, and their presence or absence (the missing dust at the beach, or the missing 100-meter craters on asteroid Eros) tell us about the formation and evolution and age of the surface. Large craters are erased and small craters are eroded by the solar wind. Boulders are cracked by high temperature swings. Every once in a while a major impact hits the "reset button" for a huge region of the surface. By studying the sizes of craters and boulders and other things, when they behave like power laws and when they do not, we can develop theories for their geology, the bombardment, and erosional processes like wind and rain.

It took centuries for the idea of planetary impact cratering to catch on. Even as modern telescopes provided detailed views of the Moon, and geologists were looking at freshly made craters here on Earth, it remained a minority opinion that lunar craters were caused by impacts. Consider the hole in the ground in Arizona that's now called Meteor Crater, the largest fresh-impact crater on Earth. Through the early 1900s, the handful of geologists who were aware of it believed it was an explosive volcano. (Coon Butte, it was called; from a distance the topographic rise of the crater rim looks like another far-off mesa. Since it is in a volcanic area, the mistake is forgivable.) Daniel Barringer, an American mining engineer of the period, reasoned that the iron meteorites that were so abundant in the area must have made this kilometer-sized hole,

Apollo 12 commander Charles Conrad stands next to the Surveyor 3 robotic lander, 200 meters from where pilot Alan Bean touched down the Lunar Module Intrepid, seen in the distance.
NASA/Alan J. Bean

and that there would be a valuable mass of iron beneath it. After much exploration an ore body was never discovered,[16] so the controversy lingered.

In 1960 the American scientist Eugene Shoemaker proved, for his PhD thesis, that this kilometer-sized hole in the ground was caused by a cosmic impact. Shoemaker's proof was inside little quartz crystals recovered from the floor and walls of the crater that he examined under a microscope. They had fine planar fractures and mineral transformations that could be achieved only by an intense shock wave passing through the rock—conditions that cannot be generated even in the most violent volcano. They require a hypervelocity phenomenon: a cosmic projectile traveling at many kilometers per second, or a nuclear explosion.

The iron fragments found in the area are pieces of what is called the Canyon Diablo meteorite, named for an expired railroad town twelve miles away. Iron meteorites were among the earliest objects to solidify in the young solar system; indeed Canyon Diablo was one of the five meteorites used by Clair Patterson in making the first accurate age determination of the Earth described above. The area around Meteor Crater has been picked over by modern humans with metal detectors; fragments up to half a ton were found historically. Large chunks of pure nickel-iron would have been a sacred novelty to Stone Age humans who might otherwise never have encountered a metallic ingot.

Coon Butte, aka Meteor Crater, was known to be an impact structure by the time the Apollo missions were gearing up, and Shoemaker had become one of the principal geologists training astronauts for the mission. Yet the geological community still remained strongly split on whether the Moon's craters were caused by impacts or volcanic eruptions. The most esteemed American geologist of the previous century, G. K. Gilbert, had concluded that the craters on the Moon were caused by impacts—but that Meteor Crater itself was a steam-driven volcanic eruption akin to those in the Pinacate volcanic field nearby in Mexico.[17] His intellectual confusion on the matter, like Hooke's, cast a long shadow, these being the great geologists of their respective ages.

In preparation for their voyages, the Apollo astronauts were trained, and their equipment and spacesuits rigorously tested, in cratered landscapes of the desert Southwest. They practiced their sorties: once on the Moon, they would have mere hours to go about an arduous series of tasks, so every activity had to be precisely rehearsed. Shoemaker's goal was to ensure that the astronauts, who were trained as pilots,[18] would know enough about geology to make precise, scientifically meaningful observations and good decisions as to which samples to bring home.

The first Apollo landings featured several-hour sorties. The later landings included an "overnight" rest and adventures on lunar buggies, and were increasingly focused on science. By Apollo 14 the astronauts were trained to do exploration geology; they layed out seismology stations and placed blast grenades that would detonate from orbit, once they were safely off the surface. They set up retroreflectors, so that scientists on Earth can bounce laser light and measure the two-way travel time, allowing the distance to be measured to millimeters. They pushed meter-long rods into dirt to measure bearing strength, and extracted cores with an electric drill. They drove lunar buggies to crater rims and along rilles, and documented what they saw. Signs of impact cratering were everywhere, in the way that signs of hydraulic erosion are everywhere in the bed of a river. They selected fragments to bring home.

The Moon's crater geology is all mixed up: the oldest, largest craters are overlain with ejecta from the next-oldest craters, like a pile of blankets on a bed that's never made. The last big ones forming on top of all that are the most prominent, overprinting everything else. With the naked eye on a dark night you can just see Tycho, a 90-kilometer diameter crater that stands out because its bright rays span the Moon, ejected by the impact of an asteroid half the size of the one that killed the dinosaurs. Apollo 16 landed inside one of the rays outside of Tycho and collected impact-melt fragments that were dated at 108 million years, so that's probably the age of Tycho.

Smaller craters form the most frequently, and LROC, the highest-resolution lunar survey camera, has detected before and after images of one swimming-pool-sized crater, made by a projectile the size of a yoga ball with the energy of a ton of TNT. Several smaller ones have also been discovered. Craters of this size show up as pits in the center of a web of subtle rays and surface disturbances. Ejected dust and rock rain down around these craters as a hail of pellets, as if a bomb has gone off, landing hundreds of crater-diameters away and sending the surface dust flying. Seismic tremors radiate away from the blast and disturb the surface dust, causing short-lived optical changes, like when a book drops onto a dusty bedcover. Somewhere on the Moon, this happens about every year.

Asteroids have craters too, and their largest can be almost as big as the asteroids themselves. One strange case is the primitive asteroid Mathilde, 60 kilometers in diameter, that is missing at least five gigantic divots (in the half that the spacecraft imaged), each of them over 20 kilometers across. It's as though it had been attacked by a giant ice-cream scoop. We've flown by only one primitive asteroid in this size range, of the thousands that are out there, so we don't in fact know whether this is unusual. Each of Mathilde's craters, caused by the impact of an asteroid a few kilometers in diameter hitting at a random angle, would have knocked it around like a piñata, leaving it spinning fairly rapidly according to any sensible calculation. But instead it is stopped dead in its tracks, spinning once every eighteen days! It's one of the slowest-rotating objects in the solar system. Either the spin from each impact just happened to cancel out—a probability of less than 1 percent—or else there is something fundamental about asteroid cratering we don't understand.

Why didn't the formation of a half dozen gigantic craters destroy Mathilde? The same was asked in the 1970s about Phobos, the moon of Mars discussed below, which is a 20-kilometer body with a 10-kilometer crater. According to everything we knew at the time, this impact should have broken Phobos apart, and Mathilde should also have been destroyed. And yet here they are.

Given all the cratering that occurs, you'd think asteroids would have to be very strong to survive these giant impacts, but the answer is actually the opposite. Imagine firing a bullet into a pile of sand and dust; it makes a crater, and you can dig through the sand and find the bullet. Now imagine that you mix this sand and dust with water, making clay that you dry into a solid brick. Fire an identical bullet into the brick, and you no longer make a crater—you make a few shattered pieces, and the bullet is also destroyed. This reasoning, backed up by computer simulations and supported by later spacecraft observations, has led to the notion of "survival of the weakest"—that asteroids have to be soft and yielding, made of porous granular materials like dust and gravel, to not be catastrophically destroyed.[19]

Mega-craters also form on full-size planets, although instead of producing bowl-shaped concavities, they damage and thin and heat the crust, and cause seismic and long-term geological disturbances that cause them to collapse into flat expanses. The largest craters on a planet can disappear entirely. The largest impacts send shock waves deep into the core, triggering global responses that can go on for days, years, or even millions of years. The largest cratering events can awaken the thermal engines inside, because if a planet is already hot, then the result is like taking the lid off a kettle: the interior starts to cool in a lopsided manner, the heat coming out of this one side where it's already thinnest. This can lead to planet-scale convection (hot stuff rising, cold stuff sinking) as the planet tries to reestablish its thermal balance. As the planet solidifies after a mega-cratering event, that imbalance can be frozen in.

Although it's difficult to measure the largest craters, that appears to be Borealis Basin on Mars. According to computer simulations,[20] a crater of this size and shape could have been formed when a wayward planet about 2,000 kilometers in diameter, a third the size of Mars, collided with a typical impact velocity and impact

South pole region of asteroid Vesta, a hugely battered asteroid whose to-pography is in fact just a bunch of impact craters and their rims, plus the 10-kilometer-wide troughs shown here circling the asteroid along the equator.
NASA/JPL/DLR

angle, with most of the projectile escaping after making the crater. But the story's undoubtedly not that simple. According to models that simulate Mars's thermal response,[21] the warm mantle would not have just sat there. It would have responded with a vigorous new cycle of convection, resulting in renewed crustal formation, creating a thick scab forming over a wound. If that is right, then Borealis is the young planetary surface, and the highlands are a second-generation crust, like a continent generated inside of the impact cavity—the exact opposite interpretation of the geology! It's important to keep an open mind in planetary science

AS A RULE of thumb, the more heavily cratered a planetary surface is, the older it is. Impacts happen everywhere, so a sparsely cratered

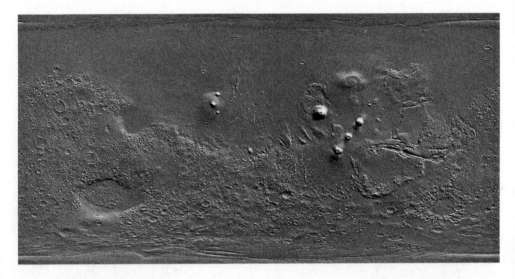

Mars, a small planet with big geology. Map view shows topography in shaded relief, from the MOLA laser altimeter on the Mars Global Surveyor mission. The northern plains, aka Vastitas Borealis, comprise a gigantic basin about 2,300 kilometers diameter. The mountains to the right, on and north of the equator, are the Tharsis Montes, the biggest volcanoes in the solar system. The east-west scar to their right (east) is Valles Marineris, the biggest canyon in the solar system. The 800-kilometer divot on the left, to the south, is Hellas Basin.
NASA/GSFC

terrain has been recently created or resurfaced by some event—a lava flow or other regional calamity, or the ongoing action of wind and water, or by plate tectonics. Using craters, we can age-date the surface. Like a wooden table that gets covered in dents and burns and becomes an antique, you can guess how many years have gone by and come up with quantitative estimates. Qualitatively, when the bombardment is mostly by small projectiles, the surface gets roughened up, beaten down, and textured. If it's mostly by big hits, then a singular feature will stand out, like the time your son set a hot pan down, burning a ring that is a marker of time gone by.

We can easily tell that the lunar highlands are much older than the sparsely cratered maria—a relative age. But it's much more challenging to come up with a quantitative age—how many billion years ago did the maria form? If an asteroid gets disrupted in near-Earth space, then after the breakup there is a storm of projectiles, and Earth's surface might "age badly" for a time. Surfaces would appear older than they are. Another complication is that we don't know quite clearly what size crater is made by a given asteroid, and this is especially true of smaller targets. Scientists have tried to come up with a surface age for the small asteroid Bennu, the 500-meter target of NASA's OSIRIS-REx mission, but to do that, we have had to guess—because we do not know—what size crater is formed by a given projectile, and what size projectile can shake down previous craters.

If you know the impact rate and the "scaling law" that gives you the size of crater that forms in a given impact, then the craters seen on a planetary body serve as a clock, provided you have a calibration of the impact rate. Here's where the Moon comes in, a body that has been mostly geologically dead for at least 4 billion years, apart from the volcanic flooding that filled the nearside basins to a depth of a few kilometers 3 billion to 3.5 billion years ago. The ages of the maria are used to calibrate crater statistics so that it can be applied throughout the inner solar system.

The craters of the lunar highlands started forming within a few million years after the formation of the Moon; the Earth was still a raw, wild place. Every new crater in the highlands destroyed previous craters; the surface is *saturated*, so it's difficult to put together a chronology beyond a certain point. At the other extreme are surfaces like those on Venus and the Earth, planets large enough to have sustained global geophysical activity. Venus is very young in that it has very few craters; it is half a billion years old. The Earth is even younger, with an average surface age of 100 million years, but with some continents serving as life rafts while plate tectonics plays out.

We don't know the process that caused or is causing the global resurfacing of Venus. If it was plate tectonics or erosion, then many of its largest, oldest craters should be ruined by the ongoing geological activity, some bordering on unrecognizable, like the oldest craters on the Earth or Mars. Instead, the largest and oldest craters on Venus are distinct and well preserved. Where are the weathered-away, partially subducted, or otherwise destroyed old craters on Venus?

It's a mystery: Venus has a very young cratering age, yet no process seems to be resurfacing it. It can't be weather: there is no water, and there's too little wind down on the sluggish surface, where the atmospheric pressure is so great. Nor can it be Earthlike plate tectonics, as this would leave ancient heavily cratered continents. Local-scale tectonics or volcanism would leave behind partially destroyed or buried major craters and impact basins. Volcanism would preferentially erase the craters in the lowlands, and would leave behind the arcs of crater rims around the flooded basins, as on the Moon. None of these would work.

Resolving these contradictions with limited data has led to the idea of the *Venus cataclysm,* that the planet's entire surface overturned half a billion years ago. Maybe the crust built up too thick, trapping in the heat, so that the whole crust foundered, turning upside down everywhere at once. Maybe there was a final impact by a Moon-sized inner planet that melted the whole surface; this is not too far-fetched. Maybe the resurfacing of Venus is related to the equally strange circumstance of Mercury, the planet that lost its mantle. Whatever happened, Venus appears to have started over as a planet, erasing any history it had.

The surface of Venus is inhospitable—it's hot enough to melt lead, the pressure could crush a submarine, like being under 900 meters of seawater, and the clouds are sulfuric acid. Nor is there any respite deeper down; the surface temperature is already greater

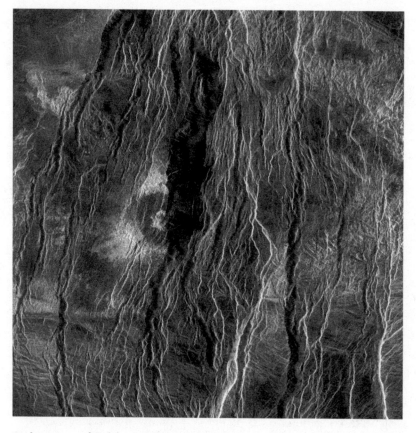

Radar image of Balch, a 40-kilometer diameter impact crater, shown in Magellan radar imaging, where bright means rough at 10-centimeter scales (the radar wavelength). Balch is one of the only large craters on Venus caught in the act of vanishing.
NASA/JPL

than any known organism can stand, and it only gets warmer with depth. If there once was life, before whatever calamity befell, it might have been wiped out. But above the clouds, 50 kilometers above the surface, the pressure and temperature are actually not far from the atmospheric conditions on the surface of the Earth. What would life hang on to there? What would it feed on?

Scientists have been able to wrestle meaningfully with these questions ever since Magellan, an American flagship mission of the mid-1980s, gave us the first geologically detailed global images, derived by processing 10-centimeter wavelength radar waves[22] that could penetrate the clouds. A decade earlier, the Russian Venera series of landed missions showed us the surface and probed the atmosphere. One hope for another flagship mission is Venera-D (*dolgozhivushaya,* "long-lasting"), whose goal is twenty-four hours of landed operations plus a radar orbiter, to which NASA might contribute an atmospheric balloon or other payload. It is the Venus explorers who have to be *dolgozhivushaya:* the anticipated launch date is always decades away, and Venus waits patiently with her secrets.

THE SUN HAS gone around the Milky Way twenty times since it was born, twenty *cosmic years.* That is also about the time it takes for the hot, solid mantle of the Earth to completely overturn, a quarter billion years, on the conveyor belt of *plate tectonics* that drives the global heat flow and is probably vital to creating truly Earthlike conditions on a planet. Originally defined by German geophysicist Alfred Wegener[23] as the breakup of the supercontinent called Pangaea—fitting the shape and geography of the continents across the Atlantic like puzzle pieces, which turns out to have been the correct idea but was lambasted as silly—plate tectonics is now understood to be a repeating global cycle[24] of shoving, grinding, submerging, and erupting.

Here's a preview of its workings. Slabs of cold and rigid *lithosphere,* also known as *plates,* sink at their margins where they get cold and heavy. They plunge into the mantle—which is hotter, more primitive, and more easily deformable—and pull down oceanic trenches. Behind those trenches, arcs of volcanoes like Japan and the Andes form in buoyant patches from the mixture of subducting slab and native mantle. Mountain ranges like the Sierra Nevada and the Himalayas are created where thicker or more complex plates converge.

Since the total surface area of the Earth doesn't change, new crust is created to replace it; mantle convection opens up *spreading centers* up and down the Atlantic and Indian and East Pacific Oceans that look like stitches in a baseball. A new inland sea is currently opening in Africa, on top of a mantle rise. Slabs today descend by convection into the mid-mantle, where they transform and dissolve; earlier, when the mantle was hotter, slabs would sink all the way to the core, piling up as a "slab graveyard."[25] Magmatic blobs from those depths might be buoyant and ascend through the mantle to erupt in hot spots like Hawai'i.

Spreading centers open up and plates collide, and in another cosmic year the continents of Earth will be as unrecognizable as Pangaea. But the underlying physics is the simple fact that the Earth is hot and space is cold, so the heat will get out. *How* it gets out forms the basis of planetary geology. Big planets have a lot more heat to get rid of than small ones, so Earth may be an optimum size for complex life, with Mars-sized bodies being too inactive and super Earths being too active. But that may be the human in me talking.

Plate tectonics on Earth has a bootstrapping problem. As the Earth's crust was solidifying, it became rigid, forming the first lithospheric plates, akin to the lunar highlands crust. But why would a plate take the plunge and start to sink? Wouldn't the crust just keep getting thicker and thicker, with heat belching out through volcanoes? A special answer is acceptable, because Earth is the only known planet with this continually transformative plate tectonic cycle. Maybe Venus is the norm, for the Earth-sized terrestrial planets that are out there.

It's been contemplated that major impacts initiated plate tectonics

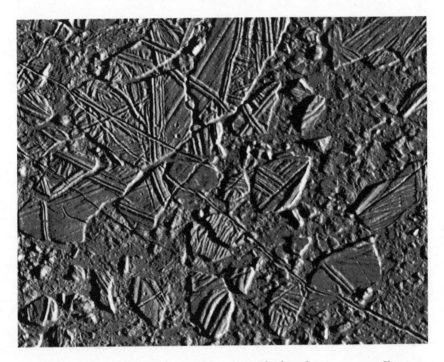

Plate tectonics of a more primitive sort. Blocks of icy crust on Europa, imaged by the Galileo mission in 1997, showing evidence for the creation of uplands and lowlands as blocks "rafted" on a warmer or liquid sub-layer some time ago (as you can tell by the craters on some of the lowlands). Elevation difference is hundreds of meters. Image scale, 35 by 50 kilometers. *NASA/JPL*

by smashing through the crust and invigorating the mantle into a global convection roll. Everyone likes to invoke a giant impact. But Venus, we think, has had equally major global impacts, and so have Mars and Mercury (proportional to their size). Maybe it had to be just the right impact at just the right time. Or maybe the Moon did it, through the incessant power of tides.

Or maybe water did it. Before plate tectonics, the surface of the Earth may have solidified into 10-kilometer-thick highlands and troughs, formed when lithospheric blocks broke apart and rafted while the magma ocean roiled beneath them, resembling the chaos regions of Europa. These troughs would have filled with the first water oceans,[26] but their topography would have been violent and unstable. At some time, for some reason, one of these plates would go down beneath another, creating a *subduction zone*. Subduction would become a pipeline for entraining wet sediments into the lower crustal and upper mantle rocks; this infusion of water would cause a wedge-shaped region to partially melt, producing buoyant granitic magma that would rise slowly and surely, forming massive *plutons* that would accrete to form the first continental crust. The cycle could begin.

Spectacular mountains like the Yosemite Valley can result when plutons rise faster than the surface weathering can erode them. But most of the action is down below, at the roots of mountain ranges where they amass into shields—thick plates as in Africa and Canada and Antarctica. When an oceanic plate collides with a continental shield, it goes under, and this generates more granitic magma that adds more plutons to the continents. When shields collide with shields, they pile up and plateaus are built like the Himalayas. The result is the topographic dichotomy of Earth: ancient continents and young ocean basins.

Venus also has highlands and lowlands, but they are not as distinctive as the Earth's terrains, and they don't seem to be the result of plate tectonics or oceans. The highlands of Ishtar Terra and Aphrodite Terra are sometimes called continents, but you can't easily draw a line around them the way you can on continents of Earth; there is no obvious boundary where they extend. On Earth, the reference boundary is sea level, with continents distinctly above and abyssal plains definitely below, with the rising and sinking of sea level by a few hundred meters defining the continental margins. On Venus and Mars, there is no continental margin, only a reference level of dry air (e.g., 1 bar or 1 millibar of pressure), and no strong geological dichotomy, either. Below we will further explore this connection: the presence of oceans makes granites possible, and granites make shields.

The oldest craters on the Earth are found on the ancient shields. The largest crater is Vredefort, a 300-kilometer diameter structure in South Africa over two billion years old that has been mostly consumed by weathering and erosion. The youngest large crater on Earth is Chicxulub, formed by a 10-kilometer asteroid or comet striking the Earth, an event responsible for the extinction of the dinosaurs almost sixty-six million years ago.[27] This impact plume acidified the oceans[28] and killed off the calcareous plankton (those with calcium-rich skeletal structures) and directly and indirectly caused the die-off of three-quarters of all plant and animal species, ending the Mesozoic era.

Until the 180-kilometer crater was discovered, buried under kilometers of sediments, all that was known was a global sedimentary record of biological devastation, and thin layers of impact ejecta in certain locales, and layers meters thick in places like Haiti.[29] More than a decade was spent searching for the crater, and there was a good chance that it would never be found—it may have been an oceanic crater that had since vanished down a subduction zone, leaving behind nothing but its ejecta[30] like the smile of the Cheshire cat. As it turns out, the asteroid struck the continental margin of the eastern tip of what is now the Yucatán Peninsula, on the boundary of a shallow sea. This explains why it was so deadly: the target rocks included sulfate-rich sediment beds, so the impact shock produced a global spray of aerosols that sharply acidified the biosphere.

The geologic record of the Moon is defined mostly by major impacts, and on the nearside, by flooding of the maria. Geologic eras of the Moon are associated with four archetypal craters. The Nectarian is named after Mare Nectaris, one of the oldest recognizable nearside basins. The Imbrian is named after Mare Imbrium, a well-

Bull's-eye. Orientale crater, imaged in 1967 by Lunar Orbiter 4 in this spectacular view from an altitude of 2,700 kilometers. The diameter of the outer ring of mountains is 930 kilometers. To the right is Oceanus Procellarum, and far to the right beyond the image is Earth.
NASA

defined basin whose ejecta are widespread all over the Moon. The Eratosthenian and Copernican eras are named after Eratosthenes and Copernicus, craters whose ejected impact melts were sampled by Apollo astronauts and have been dated absolutely. Each era is an expression of a general geologic time when craters of that size were forming.

One of the all-time favorite features on the Moon is the 930-kilometer diameter crater Orientale, the "bull's-eye basin" on the eastern limb. More than a crater, Orientale is a *multi-ring basin,* and the youngest of the major structures on the Moon. Half-hidden from the Earth, it was initially understood to be a sort of frozen-in-time tsunami, from the gigantic waves rumbling and heaving out of the partially melted, partially solid, heavily damaged crust. The newer idea, which seems to work better, is that what we see is a set of concentric scarps that formed in the crust as the pliable deep mantle surged in to fill the deep hole. Its creation and rebound would have caused a global seismic oscillation that would break down the global topography. The impact formation of Orientale would have launched massive sheets of ejecta, hurtling some of it to Earth.

Earth seems much simpler. Stand on the Pacific shore and look west, and you'll observe a wild and limitless sea. But thanks to a century of modern geology you can also see it this way: a vast slab of rock, tens of kilometers thick and thousands of kilometers across, covered in a few kilometers of water, is rumbling toward you at the rate of a growing thumbnail, a few meters in a lifetime, riding on the back of the convecting mantle. The plate takes a dive offshore and descends beneath you; where the plate is behind you, granitic plutons rise from the volcanic action triggered by the water in the slab. The unstoppable movement is met with a frictional resistance, causing earthquakes. Humans, make way.

CHAPTER 3

SYSTEMS INSIDE SYSTEMS

A great while ago the world begun,
With hey, ho, the wind and the rain . . .
—WILLIAM SHAKESPEARE, *TWELFTH NIGHT*

THE LARGEST PLANETS ARE UNIGNITED stars, rotating spheroids made up mostly of hydrogen and helium. Their atmospheres swirl with storms that can be tens of thousands of kilometers across, bigger than the Earth, driven by enormous internal heating left over from their accretion. Most of their material coagulates directly from the nebula of gas and dust that surrounds the central star, and thus giant planets are "starlike" in their overall composition. In terms of size, even a five- or ten-Jupiter-mass planet is only slightly larger than Jupiter, because these materials are compressible: the more mass you add to a giant planet, the more it compresses. There is a theoretical maximum size, a bit larger than Jupiter, at which point it simply doesn't get any bigger, it just squeezes more. If a gas giant accretes more than about seventeen Jupiter masses, then its central temperatures and pressures grow so great that hydrogen fusion begins.[1] It becomes a star.

Timing is everything in the formation of a giant planet. In what's called the *core accretion* model, the first step is the accumulation of a vast number of icy cometesimals to make a ten-Earth-mass core. If you form that core fast and in the thick of the protoplanetary nebula, then like the carny rolling the cotton candy, there is enough gravity to accrete the remainder of the disk around it, as far as it can reach. The first of these ice cores to form has all the advantages. There's more gas for it to gobble up, and the more massive it becomes, the more gravity it has to pull things in. So that big winner was Jupiter.

But although Jupiter grew at a spectacular pace, in around ten thousand years, its gravitational pull could reach only so far. A gap developed, making room for another giant planet farther out, accreting around another core of clumped cometesimals, following a similar pattern: Saturn. Farther out, Uranus and Neptune stalled as cores, according to the theory; that's why we call them ice giants. (It's a bit of a misnomer: they accreted out of ice, and their interiors are mostly water and molecular hydrogen, but now their interiors are thousands of degrees.)

If the protoplanetary disk had been twenty times more massive, Jupiter would have become a dwarf star, part of a binary system, Sol A and Sol B. Although Jupiter was luckily too small to start hydrogen fusion, there was plenty of heat left over from its accretion—the same gravitational binding energy that Kelvin considered in his calculation for the age of the Sun. (His calculation was correct, but wildly off because he didn't know about fusion.)

To understand the source of accretionary heat, consider that a giant planet forms from bits of material coming together from afar. If a shoe drops into Jupiter at the free-fall velocity of 60 kilometers per second, for example, it releases the energy equivalent to a full tank of gasoline combusting. Do that a few thousand trillion trillion times (Jupiter's mass, in shoes), and its accretion will have generated enough heat to raise its global temperature to tens of thousands of degrees. Jupiter grew in the geological blink of an eye, but its heat has been working its way out ever since, through the planet's vast radius.

Today Jupiter radiates twice as much heat as it receives from the Sun. Early on it would have glowed much hotter, baking away the water from its innermost moon, Io. From Io's perspective, Jupiter looms like a giant saucer in the sky, so it would have acted like a heat lamp. Farther out, at the distance of Europa, Jupiter would have glowed warm enough to make a steamy open sea, at least on the Jupiter-facing side, but eventually the planet would cool off and Europa's ocean would seal over—thereafter to be heated by tides and radioactive elements for billions of years. This cycle of icy satellite formation, thermal incubation, and long-term internal heating could play out wherever there are outer giant planets.

SOLAR SYSTEM CREATION begins with the falling in of gas and dust and ice to make a star and its protoplanetary disk. Planet formation begins when the disk coagulates further into solid particles embedded in the gas, like swarms of bees flying in a fog. How planets grow from there depends on many factors—the density of the disk and its composition, the radiative power of the star and its magnetic field, and what's happening nearby, interloping sister-stars and exploding supernovas. But in general terms there is a standard model. Where the protoplanetary disk is dense with particles of ice and dust, a planet can grow further like a rolling snowball, swallowing the ice and dust and eventually gathering the gas like the gluttonous character No-Face in the movie *Spirited Away*. Everything goes in.

But around the orbit of the Earth at 1 AU, the Sun's gravitational influence is powerful, and its radiation blows the gas away. Terrestrial planets have to form indirectly, not as Immanuel Kant envisioned, but in a hierarchy beginning with *planetesimals,* which accrete into *embryos,* which grow into a system of Moon- to Mars-sized bodies known as *oligarchs.* Up to that point it is an orderly and self-regulating process, the oligarchs leaving one another alone for the most part, sticking to circular orbits, dominating their local region.

Eventually, however, one oligarch runs afoul of another—a small gravitational perturbation grows, leading to a close encounter, leading to a collision, and frequently a merger—and thus begins the *late stage* of giant impacts, with planets crashing into planets. For a hundred million years the giant impacts played out, ultimately leading to the formation of Venus and the Earth, the two big winners, and beaten survivors like Mercury and fragments like the Moon. It's easy to talk about giant impacts in the abstract, but their energies are astonishing, and their physics has yet to be well understood.

We're more familiar with the process of impact cratering—planetesimals colliding with planets, as opposed to planets colliding with planets—in part because you can make a crater in the dirt in your backyard, or in your friend's tobacco-pipe factory. Also, thanks to Newtonian physics and mathematics, you can scale the formation of a laboratory-sized crater to be similar to the formation of a much larger crater by doing it in a centrifuge spun up to hundreds of g's. Small craters formed at higher gravity are *scale equivalent* to large craters. If you scale things right—the length, the time, the centrifugal force—then the two phenomena are mathematically the same.

To understand how crater scaling[2] works, consider making an action sequence for a Godzilla movie. You want the actors to wrestle and fling each other to the ground realistically. That means it has to take them a longer time to fall, being so much bigger, and it takes longer to swing their arms and legs. It turns out that two battling 54-foot monsters are mathematically similar to two battling 6-foot actors, if you scale up the time by a factor of 3 (the square root of 54/6). Shoot the scene at 60 frames per second and project it at 20, and you're good.

Crater scaling is like that, but more detailed. Let's suppose we want to understand the formation of a large impact crater on the Moon; we take high-speed images of laboratory experiments done under high centrifugal force and slow them down by thousands or millions. Once you do that—zooming in, in space, and zooming out, in time—you can directly observe planetary cratering at work.

Another approach is to use *hydrocodes,* the simulation codes that have become essential in studying planet-scale collisions—the codes that make the movies. These are benchmarked and validated against laboratory experiments. In hydrocodes, scaling is simply a matter of changing the units of time from microseconds to minutes, and distance from millimeters to kilometers, and the physics takes care of itself—if we've got the right physics. Some of the physics is easy to get right—gravity, for instance—while other physics, especially how rocks fracture and melt and flow, is harder to get right. For example, a huge pile of rock is typically weak, being riddled with flaws, while a small nugget of rock is strong, quite difficult to break. So hydrocodes can be quite complex, sometimes millions of lines of instructions, and cannot capture all the physics.

The Godzilla analogy tells us that bigger craters form more slowly than small ones, and that makes sense. Scaling in time, we see that it takes a few seconds for a large asteroid to penetrate the lunar crust, traveling at typical solar system velocities of around 10 kilometers per second. As a rule of thumb, an asteroid penetrates to a depth equivalent to twice its size before it is stopped in its tracks, unable to move any more of the planet's crust out of its way. The energy of the speeding projectile converts into shock waves that create an explosion that blasts out the cavity; in fact an asteroid impact is nearly indistinguishable from the explosion of a buried charge of the same depth and energy. The intensive process of the impact-triggered explosion scales differently from the physics that governs the final size of the crater, leading to the curious result that the larger the crater, the larger the fraction of impact melt in the final cavity. In addition, the very largest craters tap down into the mantle, which fills them with a plug of hot material, something that's been conceived of as driving hydrothermal systems on early Mars.

In the case of giant structures like Imbrium, whole mountain ranges can be flipped upside down and buried beneath masses of lower crust and upper mantle. Anything near the surface and close to the impact site escapes into space, pulverized or melted or vaporized. The initial cavity opened up by the expanding shock wave can't be sustained by the strength of the crust. It collapses and the mantle rushes in. The whole Moon shakes with a violent energy for tens of hours. The middle of the crater starts to collapse even while the outside of the cavity is still opening, so it starts to act more like a wave than a crater, oscillating like the ripples around a rock thrown into a pond.

For craters bigger than a few hundred kilometers, you can no longer pretend that the Moon is flat. The South Pole–Aitken basin (SPA) and Oceanus Procellarum are gigantic structures that each wraps a quarter around the Moon. The curvature of the Moon matters, as does its crust and mantle boundary, and the fact that "down" points in a very different direction from one side of the crater to the other. Crater scaling becomes highly uncertain for these largest events, so provides just a rule of thumb; here our best hope is to simulate them realistically with hydrocodes tuned to the geometry of a spherical planet in space.

Giant impacts are even more complex. There is no center to such a collision, and there is no laboratory where you can play out a scaled-down version of the event. Still, we can make some estimates. Giant impacts, caused by planets getting entangled gravitationally, occur at around the *escape velocity* of two colliding bodies—the speed at which one body falls onto another in the ideal situation where they are the only objects in the universe, 11.2 kilometers per second for the Earth and 2.4 kilometers per second for the Moon. This makes them the "slowest" collisions, when you think of how they scale: if the projectile and target are comparable in size, and they collide at around their escape velocity, then it takes an hour just to finish the physical contact. After that, the impact plays out for days, the gravitational and mechanical interaction of two bodies whose cores are merging, or trying to—this is something that we can study only with 3D computer simulations that solve for the gravitational attraction of all the material, interacting with all the other material, and that includes all the other physics of impact cratering. It starts to look like interacting galaxies, and indeed, the same kinds of hydrocodes are used.

ONE ASPECT OF collisional physics is easy. If you want to accrete two planets into one place, then you also have to accrete their momentum; that's Newton's law. When you fire a pellet deep into a block of wood, the block accretes the momentum of the pellet, so it flies off the table at a speed equal to the speed of the bullet divided by the relative mass of the block. The same thing is true for rotational momentum. Imagine an astronaut floating in space who grabs a spinning equipment box. She ends up rotating but at a slower rate, proportional to her larger moment of inertia.[3] If the box is spinning too fast and is too massive, she can't hold on.

That inability to hold on is quite analogous to two planets trying to grow by colliding in a giant impact. Because of angular momentum, it is hard for two similar-sized planets to grab on to each other, and we have observed in computer simulations that giant impacts lead to planetary mergers only half the time. The other half turn out to be *hit-and-run collisions*, where the two planets continue their separate ways, dazed and battered.

Sometimes the smaller escaping planet, the *runner*, survives in a sense but is torn into multiple escaping sub-planets, emerging from the collision like beads on a string—a mechanism of forming families of planetary bodies that have a common chemistry. They were scooped from the same "stew pot," although some of them will have all the meat and others will have only potatoes. Off they

Hubble Space Telescope image of comet Shoemaker-Levy 9, which was torn apart by tidal forces in a near miss with Jupiter in 1992. Image taken in 1994 a few months before the fragments collided with Jupiter on their next pass. Hit-and-run collisions in the early solar system would have produced similar structures, "beads on a string" of planet-sized bodies.
NASA/HST

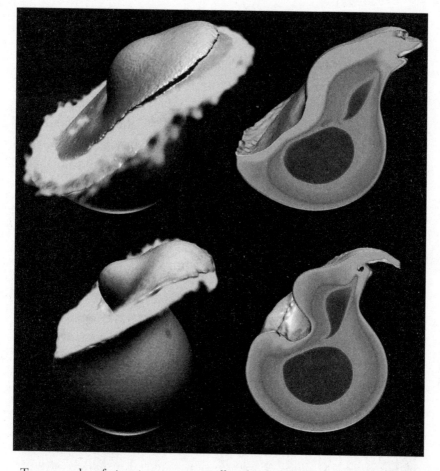

Two examples of giant impacts, one will make the Moon, and the other will not. The colliding bodies are shown one hour after making contact. To the right is the cross section, plotting density, showing the cores sailing past each other in the top case (a hit-and-run collision), and about to become entangled in the bottom case.

Simulations and visualizations by Alexandre Emsenhuber (U. Arizona)

would go, into the wild blue yonder, half a dozen new planets, some of them rock rich, others metallic, some oceanic . . . It's not just speculation, because we've seen an example of such an event in 1992, when Jupiter pulled a comet into pieces.

The standard giant impact model of Moon formation is halfway between an accretion event and a hit-and-run event. An oligarch one-tenth the mass of the Earth, the size of Mars, collides in one of the last great sweepups, at an impact angle of 45 degrees[4] at close to the escape velocity—quite typical, from what we understand of the late stage. Being more than half the diameter of Earth, Theia didn't make good contact; even though it's the most probable impact angle, it's not the most probable accretion angle. For impact angles greater than 60 degrees (about a quarter of all impacts), almost every giant impact is a hit-and-run. You need a relatively head-on collision angle to stop the momentum. Theia was slow enough when it collided that it lost enough momentum to be captured gravitationally by the Earth, shredded apart by the collision and by Earth's gravity acting as a huge tidal force. Its shredded body looped back around ten hours later for a second go, this time slower, resulting in accretion.

The cores merged rapidly in the standard model, by which I mean a few hours. That is the gravitational time scale of the system, $(G\rho)^{-1/2}$ where G is Newton's gravitational constant[5] and ρ is the matter's density, grams per cubic centimeter (water = 1, rock = 3). The cores sink and merge, and that accumulates enormous spin. The core-mantle boundary gets dragged along for the ride, and the planet flings off globs of mantle-rich material just like Darwin envisioned; in fact you can think of it as impact-triggered fission. The core of Theia is almost completely accreted by the Earth, so the Moon ends up forming out of a protolunar disk with a rocky composition.

If Theia had been traveling just 10 percent faster, then it would have emerged as its own planet, completely devastated, but escaping the Earth and going back to orbiting the Sun. But the collision would have slowed it down, so it would be nearby and likely to collide again someday. When there is a series of bounces it's called a *collision chain,* with each one slowing down the oligarch so that the next one is more likely a merger, and if not that one, the next, like a Slinky going down a few carpeted steps.[6]

GIANT PLANETS ARE known for their remarkable atmospheres and interiors, streaked with clouds and swirling maelstroms. Yet however mesmerizing their appearance and no matter how dominant they are in dictating planetary dynamics, giant planets don't have *surfaces*, as is the case of the terrestrial planets and the large icy bodies like Pluto and Europa. A surface is a two-dimensional boundary between liquid and gas at the surface of an ocean, or between air and land, or between water and ice. Surfaces are boundaries between phases of matter that provide material contrasts and niches. It is a wild, wild world out there, subject to violent whims, and surfaces offer places to call home, to find shelter and warmth, a place to rest.

There are organisms on Earth that do not live on or near a surface: the free-floating plankton of the sea and the sleepless sharks. Yet these are ultimately dependent on what goes on at the ocean surface—the exchange of O_2 and CO_2 with air, the input of solar energy, the primary productivity of organisms living in this sunlit oxygenated layer. Creatures that never see the Sun feed on near-surface creatures (on Earth there is a daily migration from below, to graze on the upper strata of phytoplankton at night). Deeper still, where organisms sink when they decay, is the bottom surface, where stranger organisms thrive around the geothermal vents and black smokers. The most abundant and diverse life on Earth is found among the gradients and contrasts of the intertidal zones, creating a land-air-sea interface covered and uncovered by the gravitational pull of the Moon and the Sun. The crenulation and porosity and permeability are where chemistry can happen, supporting the thermodynamics of life, exchanging water and solutes across membranes, through alveoli and capillaries and gills.

One mantra for planetary exploration is to follow the water. We should also follow the *niches*. Here water is again unique, being able to create its own nooks and crannies because of its behavior around the triple point, where gases and liquids become solids (freezing), solids become liquids (melting or dissolving), and solids and liquids become gases (sublimating, evaporating). Furthermore, all kinds of molecules dissolve in water, where they can dissociate, recombine, and precipitate into new solids, and change the physical behavior (e.g., density) and chemistry.[7] Most of Earth's freshwater is found in subterranean aquifers in the upper crust. At the highest latitudes north and south, freshwater is stored in ice sheets up to several kilometers thick, and these regulate the Earth's climate by reflecting a fraction of the Sun's incoming energy to space, behaving like the window shades on your car. Beneath the thickest ice sheets, the pressure is so great that geothermal heating causes melting, so there is liquid down there as well.

Ice sheets extend out onto the ocean as permanent and seasonal ice shelves, as a result of the remarkable fact that solid water floats on liquid water. At a couple of times in Earth's geologic history, the period of the *snowball earth*, these ice shelves extended down to the equator. You would not have recognized our planet from space back then; it may have looked like Ganymede. In the snowball earth, most of the Earth was an ice shelf, analogous to what is found beneath the cryosphere of Europa. Ice shelves on Earth are labyrinths of cracks and ridges, terraces and caves. They are havens for amphipods, algae, and polar cod. The ice sheet facing the water is a lot like the seafloor, only upside down, a niche-rich environment that might be common on icy worlds throughout the galaxy.

Earth's hydrologic cycle is an incredible perpetual machine, marvel of the solar system, life's blessing. Consider just this one aspect of it: runoff from the continents by rain. Flowing water brings dissolved minerals to the ocean; ions from these minerals react with atmospheric CO_2 dissolved in the upper meters. Calcium dissolved in the water makes carbonates as described below, which pull CO_2 from the atmosphere. If it gets too hot (too much atmospheric CO_2), this causes more rain, which leads to more runoff. This brings more calcium to the ocean, which means there is a further drawdown of CO_2, leading to climate cooling.

Somebody gave us a magnificent self-regulating machine. But we're like a guy with a pocketknife tinkering with a Swiss watch, messing up the most important feedbacks—the loss of the reflective ice sheets, which means less of a sunshade, and the increase in methane, the potent greenhouse gas being released from the permafrost as it thaws. Planets breathe in long rhythms, and sometimes they sneeze.

ICE MELTS UNDER pressure for the same reason that it floats above water: the solid state takes up more volume (less dense) than the liquid state. So if you force ice into a smaller volume by applying pressure, you change it from solid to liquid without changing the temperature. That's why a skater's thin blade glides without friction: all his weight is focused onto a small cross section contacting the ice, so the pressure is concentrated there and the solid turns to liquid for a moment—and solidifies behind the blade.

The high latitudes of Mars are dominated by ice-cemented soils, and toward the poles there are the ice caps. Peering 1.5 kilometers below the surface, radars have detected echoes from a shallow aquifer that some have called a "lake." Its presence is not too surprising, given that liquid water can exist at depth even if the surface is frozen. The deeper you go, the hotter the temperatures and greater the pressures; there will be a zone where brine can accumulate and remain unfrozen. Although it is probably toxic to the organisms we know and love, this might be where original life on Mars evolved, as the surface became cold and harsh and inhospitable. Are there microorganisms down there, adapted to a pickled existence? The first step in Mars subterranean exploration is happening now; as I write this, NASA's InSight lander is attempting to drill 5 meters in, but finding it difficult. Drilling a thousand times deeper will require a Mars infrastructure we haven't imagined.

Then there's the transition of solid to vapor, known as *sublimation*. Ice cubes deep in the freezer vanish, leaving behind cloudy nuggets that nobody should put into their drink. What happens to the H_2O lost from ice cubes and from mummified fish sticks and forgotten string beans? It gets deposited in intricate solid form as frost on the freezer walls and accumulates into coarser, larger grains. This sort of experiment happening in a dorm-room freezer is actually not a bad analogue to the processes going on at the surface of the ice-rich soils of Mars, and on comets and icy moons, although to do the experiment right, you need to put your freezer into near-vacuum conditions and crank it colder and expose everything to ultraviolet light. There are about a dozen planetary science labs around the world doing just that.[8]

Each phase of water is a pattern or *state* in which molecules of H-O-H can be arranged: locked into crystalline lattices (ices of various structure[9]), free to move around (the liquid and gas phases), or solid but not arranged in lattices (amorphous solids). There are numerous other phases of H_2O, more than a dozen so far discovered (and yes, there is *ice-nine*, but not nearly as interesting as Kurt Vonnegut[10] said it would be). Half of them are relevant to geology.

Because planetary water is never pure, these phases are idealizations. Add a lot of salt and you have a brine that is denser than water and remains liquid at a much lower temperature. This is why people put salt on winter sidewalks; it lowers the melting point, making it harder for water to freeze. A planet made of briny H_2O can have a very intricate geology; Tethys is one such body; Iapetus is another, with its "walnut ridge," which some think may be the result of being cracked open by expansion stresses when water freezes. (Others think it is the result of ice-brine convection. Still others think it is the result of a sub-satellite that once orbited Iapetus, which crashed into it. Yes, Iapetus is weird.)

At deeper pressures, especially under added heat, silicate mineral structures can rearrange to accommodate water, creating *hydrated silicates*, like gypsum and serpentine, where water is linked into the molecular structure, or by becoming clays. Hydrated silicates may dominate the mantles of the largest satellites—Ganymede and Titan—mixed by thermal convection or overturned by the giant collisions of their accretion. Liquid water in intimate contact with warm rocks, and enough geothermal energy to drive global-scale convection—is that enough for life? Sunlight would be nice, but we know from the existence of deep subterranean organisms on Earth that it is not essential.

The presence of liquid water at the surface-atmosphere interface (that is, an open ocean) acts as a buffer on planetary temperature. This keeps Earth at around the conditions of the triple point. So long as there are oceans, a planet can't get hotter than boiling, because any energy gets used to convert water to steam. (Once the water all evaporates, all bets are off; look at Venus.) Nor, as long as the open ocean remains unfrozen, can it get colder than freezing, on a global scale.[11] The presence of humidity in the air also buffers temperature, and our skin emits humidity into dry air to keep us cool.[12]

If this liquid/vapor interface is gone (e.g., no ocean), then these buffers disappear, and temperatures can rise and fall freely. When the temperature begins to plummet on a cold evening when it is damp, then at some point it stops getting colder (though it feels quite cold, being damp). Vapor in the air starts condensing on the wet branches and the dewy grass and on your clothes, and it may even start to drizzle. These are droplets, aerosols, tiny airborne condensates with surfaces. The temperature stops dropping while the dewdrops form and while it drizzles, according to a balance of a quantity known as *entropy* (which I won't try to explain because I don't understand it). In the dry desert, where the humidity might hover at around 20 percent, the temperature on a warm sunny day can plummet to below freezing without a hitch because there isn't any water to condense. Also, because there are fewer H_2O molecules to absorb the infrared radiation, the heat gets out fast, so it is cold by nightfall. Not so in the tropical jungle or on a CO_2-rich swamp planet.

As a last familiar example, a jar of water in the freezer drops to 0°C and then stays at that temperature as long as there is any liquid left unfrozen. Once it has solidified, the ice cools further. In the process of solidifying, the contents will have expanded by 9 percent, so if the jar is too full it will break. The expansion of ice does more than break jars and make water float to the surface in oceans and lakes; it is also one of the major erosional powers on Earth. H_2O works its way into rocks, then freezes in place, and its expansion cracks rocks apart; this is how they used to quarry Italian marbles, by pouring water into slots they would carve into the rock in the winter. The water would solidify and expand, exerting a huge and relatively even force. Cyclical freezing-expansion and the fact that water molecules are attracted to other water molecules (because they are polar—that is, asymmetrically charged) cause all kinds of geological upheaval, from the emergence of boulders wherever there is a deep freeze–thaw cycle in the soil,[13] to the sprouting of pingos in regions of the arctic—masses of ice up to hundreds of meters diameter that get pushed out of the ground like a zit due to the accretion of water around a permanently frozen core.

EARTH HAS BEEN called the Goldilocks[14] planet: not too hot, not too cold.[15] It orbits in the habitable zone of a stable star, and has a 1 bar atmosphere[16] made mostly of nitrogen (78 percent), oxygen (21 percent), almost 1 percent argon, and 410 parts per million CO_2, plus a varying fraction of H_2O vapor that depends on temperature and pressure and location, typically around 1 percent at sea level. Without this atmospheric blanket the average surface temperature would range wildly, averaging zero Fahrenheit (−18°C) with enormous oscillations. The oceans would solidify.

The trace gas CO_2 has increased by more than 40 percent since the beginning of the industrial era, as we mine long-sequestered fossilized carbon (oil, gas, coal) to run our engines. Carbon dioxide, the major product of internal combustion, is an invisible gas in the sense that visible wavelengths of light don't interact with it, and it is largely inert. It doesn't bother us unless it displaces the oxygen we breathe. Plants thrive on it, photosynthesizing CO_2 and H_2O, ultimately producing atmospheric oxygen (O_2) along with the organic compounds of the biome and the energy sources (sugars) necessary for carbon-based plant and animal life.[17]

CO_2 is a *greenhouse gas,* and now we have too much of it. The Sun's visible (yellow) rays go right through it to warm the land surface and the ocean. But the outgoing heat from the warmed surface, in the infrared, can't get out. This is the same reason why the sky is opaque to infrared astronomy.[18] Actually, H_2O is the most potent greenhouse gas of all, but its contribution to warming is secondary: when it is cold, the air is dry, so H_2O has no effect. When it is warm, the atmosphere holds more vapor, which amplifies the greenhouse warming.

On Venus, which was born closer to the Sun and has an atmosphere of primarily CO_2, the greenhouse atmosphere was a runaway disaster. The hotter Venus got, the more of its water evaporated, making the planet warmer and causing water to escape with the strong solar wind. (Solar radiation interacts more intensely with Venus than with Earth because it is closer, and because it lacks a magnetic field.) After billions of years, most of its original water was gone, although significant water might remain buried in the mantle waiting to get out. Maybe it didn't have to be that way. If only there had been a geologic process that could have removed the CO_2, maybe it would not have gotten so hot in the first place, and there would still be liquid oceans on Venus. The irony is that if there had been liquid oceans on the surface, maybe there would have been a geologic process that would get rid of all that CO_2.

Near the ocean surface of the Earth, water is aerated by the turbulent action of waves and the spray of foam. Fish breathe the dissolved air through their gills. The atmospheric CO_2 also dissolves, and some of it reacts with water to form carbonic acid,[19] $H_2O + CO_2 = H_2CO_3$ (to simplify a longer series of reactions). At the same time, calcium ions (Ca) are eroded from the continents into rivers that flow to the sea, and these react with the acid to release hydrogen ions and make *calcium carbonate*, $CaCO_3$. One major sink for CO_2 is biological, the fossilized skeletons of corals, forming a rock called aragonite, and the structures of plants and algae and plankton all over the sunlit land and ocean surface.

But the greatest sink for atmospheric CO_2 is precipitation into solid form, when the water chemistry becomes too rich with Ca and CO_2, like when your teakettle gets covered in mineral deposits.[20] Crystals of $CaCO_3$ accumulate on the seafloor,[21] and when those deposits get thick enough they are compressed into rocks like dolomite and limestone. Eons later these can get pushed up by the buckling and faulting of plates, exposing fantastic geologic structures like the White Cliffs of Dover and the Stone Forest of Guangxi.

This drawdown of CO_2 by the production of carbonates in the ocean can work as a pump only if you can keep it going. Here is where plate tectonics comes in: on a time scale of the order of a hundred million years, the seafloor is subducted into the mantle like a conveyor belt, bringing the carbonate-rich sediments down with it. Early on, this would have caused the drawdown of a CO_2-rich atmosphere.

Once the cycle of plate tectonics got going, a fraction of the buried CO_2 would start to come up again in volcanic vents, so today there is an equilibrium, where most of the CO_2 is plowed back into the crust and mantle. If you add up all the Cliffs of Dover and extrapolate globally, and add what you think is dissolved in the mantle, this represents the disposal of around 10 bars of CO_2 atmosphere, beneath which the Earth would otherwise swelter.

Suppose you were able to precipitate all the carbon dioxide out of Venus's atmosphere by this process of continents weathering into global oceans and precipitating into carbonates. The result would be an outer carbonate crust 800 meters thick[22]—a brightly colored surface that could reflect the Sun's light into space. The result could be a marginally habitable planet—and a spectacular landscape, although uncomfortably stark. You would want good sunglasses.

In the immediate present, here on Earth, "terraforming" is sadly no longer science fiction. We have decided as a species to unbalance the biogeospheric system of carbon sequestration for our short-term convenience. Last I checked gasoline was under $3 a gallon, so cheap as to speak of reckless use of buried carbon despite immediate common facts.[23]

WATER IS THE solute for life as we know it, but its variability is tremendous. The details are lost, but we know that life started on

Earth with *extremophiles*, who were content to live in the aqueous conditions of the late Hadean. These included the halophiles, thermophiles, and barophiles—organisms that thrived in brines and near geysers and at depth. They're not gone—they've moved to the niches (as we have speculated about the deep briny pockets in Mars). But every last one of them, no matter how extreme, requires water at some phase of its existence. So how about elsewhere?

One obvious place to consider is 5,000-kilometer diameter Ganymede, the most massive ice-covered body in the solar system. It is the largest known satellite, until we detect giant

Mosaic of seventeen images from NASA's Voyager 1 flyby of Ganymede in 1979 shows resurfaced rifts, angular blocks, fresh pockmarks, a few large craters, and no lunar-like basins. Beneath this patterned surface, under 100 kilometers of ice, is a liquid briny ocean transitioning to a basement layer of ice VI, and then hydrosilicates, and finally a rocky mantle and dense metallic core.
NASA/JPL

moons around planets around other stars. Ganymede has an internal ocean deep inside, known from magnetic field measurements by NASA's Galileo mission (discussed below) and the simple fact that ice melts under heat and pressure. The depth and extent of Ganymede's oceans are unknown. Its chemistry and mineralogy must be inferred, and the geology beneath the ice is a wild guess.

We know it is covered by a rigid ice shell 50 to 100 kilometers thick, and there has been no oceanic interaction with the surface since the last major collision billions of years ago. It will be more challenging to send a robot down into the strange seas of Ganymede than to send a mission to the nearest star system, but in the meantime we can look, and think, and explore.

Ganymede's oceans are heated in a number of ways. First there is the gravitational binding energy from when the body came together, which is sufficient to melt it completely during impact accretion. This accumulation was relatively rapid, but the heat inside might take hundreds of millions of years to get out. Next is heating from the decay of uranium and other radioactive elements contained in Ganymede's rocks; some decay fast but others are long-lived like potassium, thorium, and uranium, producing energy for billions of years and contributing most of Ganymede's internal heating. Then there is tidal heating, which is also substantial, and for certain satellite systems is the longest lived of all, conceivably cranking out heat for a trillion years.

To understand tidal heating, let's first consider the Moon, whose nearside is tidally locked to the Earth. If the Moon's orbit was perfectly circular, then the Earth wouldn't appear to move in the sky, seen from anywhere on the nearside, although the Sun and stars would rise and set each month, which is also its day. But the Moon's orbit is not quite a circle; it is an ellipse with 5 percent eccentricity. Obeying Kepler's law, the Moon orbits a bit faster when it is closer, at perigee, and slower when it is farther from the Earth, at apogee. Because the Moon's spin remains constant, it orbits faster than it spins during perigee, and spins faster than it orbits during apogee. So if you're living in a vacation condo near Orientale Basin, 90° west longitude on the Moon, the Earth will always be on the eastern horizon, rising a little bit and setting a little bit each month—a very pretty view. The Earth, circling around like that, raises oscillating tides inside the Moon that cause friction, generating heat, like bending a paper clip repeatedly.

Today the Moon is solid and elastic, the Earth is distant, and the eccentric wobble is only slight, so tidal friction generates maybe enough heat to contribute to what might be a solid-liquid mush surrounding the small iron core. But early on, when the Moon was a great deal closer to the Earth, tidal heating was enormous. Jupiter also makes little circles in the sky, seen from Ganymede's surface, but more frequently (every seven days); plus it exerts a more powerful tidal force, generating significantly more heat. Ganymede's mushy oceanic mantle shears back and forth, and although it would not be enough heating to melt Ganymede today, if it was completely frozen, it's enough to prevent its briny ocean from ever solidifying.

Nothing comes for free, and tidal friction takes energy out of a moon's orbit, making it more circular with time,[24] but in the Moon's case, at a rate so low that we can infer that it is a solid elastic body.[25] If there were no other factors, then Ganymede's tidal heating would eventually shut down, Jupiter would stand still in the sky, and the tidal bulge would become permanent. But Ganymede is not the only satellite of Jupiter; it is locked in a *mean motion resonance* with two other Galilean satellites, Io and Europa. As we shall see, this means that Ganymede has a *forced eccentricity* determined by the mutual gravitational interactions with the other satellites, a circumstance that can be maintained for many billions of years, so the tidal heating never stops.

We know from basic exoplanetary data that there are super Earths out there with open oceans five to ten times the surface area of Earth, tens of kilometers deep. One can only imagine the currents and the storms, the power of tsunamis generated by super earthquakes. Could mountains and volcanoes rise up from the seafloor of a water world to make islands and continents? I think not.

Mountains can rise up only so far before they collapse under their own weight, and that's as true underwater as it is on land, especially on a massive planet with more gravity. A more serious complication is that H_2O transitions to a solid phase called ice VI at high pressure. Earth's oceans don't get deep enough to turn into ice VI, but Ganymede's oceans extend hundreds of kilometers deep and may solidify.[26] A super Earth with oceans more than 30 to 40 kilometers deep would (extrapolating from the laboratory data) have a seafloor made of ice VI, breached by volcanic eruptions.

Water is not entirely transparent to light, so the primary biosphere of a sunny water world would be in the upper tens of meters, if it existed, teeming with plankton or colonized by vast microbial mats. At the bottom of this ocean, above the basement crust of ice VI, life might huddle in utter darkness around vast volcanic regions, feeding on material coming up through the vents (black smokers) and on decaying organisms drifting down from the surface that decompose into a high-pressure ooze. This is all conjecture, of course, and probably bad conjecture, but we live in the world of *who knows?* The water worlds discovered so far are quite close to their star and likely to be shrouded in steam, and therefore less interesting from the point of view of life. But planets farther away from their star are harder to discover, so "Goldilocks water worlds" are out there, perhaps already found. And then, even farther from their star, the most distant water worlds are ice worlds, although if super-Earth-sized, their cryospheres would be geologically active from the substantial internal heat needing to get out, the time scale of cooling for such a massive planet being billions of years.[27] Imagine the geology of such a planet, if you can.

Water on a terrestrial planet contains minerals dissolved by interaction with the silicate crust, and from the mantle. Add magnesium and sulfur and sodium and chlorine and ammonia—the components of salts, that is—and you have a brine that is denser than pure water and is liquid at temperatures that would solidify pure water. When a briny ocean starts to freeze, the first crystals to form are composed of freshwater. These freshwater crystals float, producing an ice shelf. This situation—a brine that resists freezing, capped by an insulating ice shell—is a recipe for long-lived liquid water oceans throughout the galaxy.

As the ice shell thickens, the residual water gets brinier and brinier, becoming a weird substance that can remain unfrozen down to −30°C to −60°C, but becoming viscous, like a lava. That's what makes Pluto such a cryo-fantastic world, the weird and unexpected behavior of ices and brines. Some of the most fascinating geology in the solar system is from *cryovolcanism*, which happens when a briny aquifer or ocean freezes its last liquid. The expansion of solidifying ice causes the briny residue to squeeze out under pressure, confined into a smaller and smaller volume, like a hole in the toothpaste tube. It gets extruded as an eruption, flow, or viscous plug, and can also cause global expansion and regional fracturing of an icy planet's crust, of which there is indication in the upper hundred kilometers of Ganymede.

The next exploration of Jupiter's icy moons will be NASA's Europa Clipper, to launch in the 2020s and make repeated flybys of Europa. Its radar is designed to detect the ocean at the base of the ice sheet that floats on the ocean. ESA's JUICE mission[28] is being developed to make flybys of the Galilean satellites in 2030 and is designed to end with an eight-month orbital tour of Ganymede. Its radar will be able to peer tens of kilometers deep and may find subterranean lakes, maybe on a vast scale similar to the chains of lakes around Vostok in the Antarctic of Earth. But truly understanding the interior of icy moons will require more than orbital "remote sensing" exploration. This will require a set of landers distributed around the surface, each one collecting seismology data that would be knitted together to construct a 3D seismic image.

That's in the far future. The surface is treacherous with radiation, crystalline wonderlands full of cracks and pits and penitentes— places that make landing engineers wake up in a cold sweat. For now, the subsurface oceans can be explored only in our minds. Any conceptions we may have are based around laboratory experiments and our experience of diverse and extreme suboceanic and aqueous environments on Earth: Vostok, the Marianas Trench, the submerged caverns of Quintana Roo.

PLANETS WITH SURFACES are a special category. You can land on them, coming to rest on an impenetrable barrier of solid ice or rock, or an open ocean. Titan is among the most benign of all the planets there are to land on, as it has a massive atmosphere that you can plunge into, just as astronauts do atmospheric reentry on Earth, but coming in much slower. It will be an excellent place to fly around in someday, with its dense, stable atmosphere and low gravity.

If you were to free-fall onto a gas giant like Jupiter or Saturn, what would happen? First, you'd be accelerated by tens of meters per second for every second that you fell, kilometers per second faster every minute. A quiet but spectacular ride going in, you'd eventually hit the tenuous upper atmosphere at 60 kilometers per second, in the case of Jupiter. It would build up a ram pressure and a turbulence that would generate a vibration more intense than any vertebrate could withstand; hopefully you have a vibration isolation chamber. At five times the speed of an astronaut coming home to Earth, your capsule would have 5 squared or twenty-five times the kinetic energy, and this needs to be dissipated as heat, requiring a thermal ablation shield that is twenty-five times as massive as Apollo's.[29] Supposing you lived, you would then gloriously descend by parachute through clear Jovian skies; below you would be an amazing multicolored cloudscape. You'd struggle to enjoy it, however, because in Jupiter's gravity you'd weigh a quarter of a ton.

Atmospheric pressure would increase as you drifted down toward the clouds, your vessel creaking as its hull adjusts from the vacuum of space and hypersonic entry, to this slow descent into the depths of the planet. Soon you'd get to 1 bar pressure, like the surface of the Earth. Deeper, at 2 bars, Jupiter's atmosphere is room temperature, and although the pressure is what a scuba diver experiences 10 meters deep, it's bearable. But you wouldn't want to open the door: the air that would rush in would be a dense and toxic mix of hydrogen, sulfur dioxide, ammonia, and methane. And with winds and gusts up to hundreds of knots, you'd want to keep the hatches battened.

If the powerful winds shredded your parachute, you'd drop like a stone. You would feel an initial relief—the burden of gravity would be lifted, as you'd be in free fall. The last part of the story is short: you'd punch through the cloud deck in a final hurrah, and then be slowed down again when you hit the denser layers, and ultimately crushed.[30] We know something of the deep atmospheric structure of Jupiter from the Galileo probe, which was deployed from the spacecraft at Jupiter in 1996. The probe's final communication was from a depth of 160 kilometers and a pressure of 22 bars and a temperature of 152°C. The rest is conjecture: its parachute would have melted shortly thereafter, and within an hour it would have sunk to the supercritical pressures of fluid[31] hydrogen, where its titanium hull would dissolve.

> As the reader will observe, there is nothing
> problematical about this deduction whatever.
> —PERCIVAL LOWELL, *MARS* (LOWELL OBSERVATORY, 1895)

The most familiar planetary surface besides the Earth's is the one that maverick American astronomer Percival Lowell said was flowing with canals dug by a waning race of giants, to irrigate their oases across the Martian desert: "His muscles, having length, breadth, and thickness, would all be twenty-seven times as effective as ours. He would prove twenty-seven times as strong as we," gravity being a third as much. Physics aside—the atmosphere is too thin and the planet too cold for this sort of thing—Lowell's books stirred up a popular interest in planetary science befitting his era.

Carl Sagan, a more modern but equally open-minded advocate for the search for alien life, studied the chemical evolution of pre-biotic compounds on Mars and also Titan, Venus, and the early Earth, and on comets and asteroids, and throughout the cosmos. He recognized that our planets are a few drops in the bucket of what's out there, and that we have to be lucky in figuring out where to look; he was also a powerful advocate for two of the greatest missions of exploration ever, Voyager and Viking.

The first fully successful landers on the Martian surface, Viking 1 and 2 were launched in 1975, and operated almost flawlessly for several years. Each mission consisted of a pair of spacecraft, an orbiter to map the surface, plus a large lander to assess habitability and do experiments to see if anything was alive. They set down in the frozen deserts of Chryse and Utopia Planitia and made it through repeated nights and frosted winters, comfy with their nuclear power packs. These were amazing missions, but ultimately they were unlucky when it came to finding life or even the plausible conditions for it. Being fixed landers, they were unable to move to a new location. The debate turned to whether the experiments were the right ones or were done in the right places, or neither.

Science is fickle, and as the Viking missions faded, it turned its back on Mars. Energies for exploration went elsewhere—the Voyager and Galileo missions blazed through the outer solar system, Magellan mapped the geology of Venus, and the Russians made significant attempts to land on the Martian satellite Phobos. Japan, Europe, and Russia flew missions to intercept Comet Halley in 1986. There would be no further successful missions to Mars, by NASA or any other agency, until 1997. Incredibly, there were no missions to the Moon, either, between 1976 and 1990.

Without further data, arguments about Martian water became abstract and contentious. By 1988, when I entered grad school and was first exposed to these modern ideas, scientists were divided about whether Mars even *had* evidence for substantial liquid water. One mainstream idea was that the channels and canyons were carved by windblown sand acting for billions of years, or by dust-laden carbon dioxide discharged from the ground, flowing like rivers—anything but liquid water, which would be extraordinary. In hindsight, this skepticism in the 1980s about water on Mars was odd, even though skepticism is what science is best at. Since then, thanks to modern landers and orbiters with advanced spectrometers and radars, the evidence for liquid water has become irrefutable. It has flowed through chasms and canyons, and some lazy-flowing meandering rivers, and filled up wide crater lakes. Water flooded the northern plains, the controversial Oceanus Borealis. The scientific rebound has been so strong that Mars exploration now holds the lion's share of NASA's solar system exploration budget.

Most ambitious of these plans is a sample return campaign that will begin with Mars 2020, a massive rover that will land near a river delta inside of Jezero crater and collect a cache of samples. If all this goes well, next will be a mission to land right next to this cache, pick it up, and launch it into orbit around Mars (somewhat how the Russian Luna missions in the early 1970s launched samples from the surface of the Moon, but now using a larger return rocket to match the higher escape velocity of Mars. Last would come the grab-and-go mission that would retrieve the cache from orbit and return it to Earth. It all sounds quite complicated, and it is. These three missions combined will have about the same $10 billion price tag as JWST, so the project has to be subject to the same level of scrutiny. (Mind you, $10 billion as a number is really not that big.[32]) Call me a skeptic, but the eggs are all in one basket and the risks are high. We have never launched a rocket from the surface of Mars, and we have never retrieved a payload from orbit in deep space, where the time lag in communications with Earth ranges from five to twenty minutes. And if the first two components go well, the last mission will become too big to fail, and grow extraordinarily in costs. I suggest instead that we develop advanced robotic exploration of the Moon first, thereby raising the technological readiness level (TRL), and then directly pick up the Mars 2020 caches that will be left at Jezero, perhaps even on schedule.

If the Mars sample return were to bring back unequivocal evidence for life, should we be surprised? *Of course* life existed on Mars, around four billion years ago. We know this deductively. Life was flourishing on Earth in sufficient abundance to have left an imprint on the rock record around that time, and surface materials were swapping back and forth between neighboring planets, an epoch of *ballistic panspermia,* where hardy biota (for example, spores, viruses, dormant bacteria) rode as passengers on large rocks ejected into space. Craters the size of Imbrium and Orientale were forming on the Earth, and life was flourishing. Large masses of upper crustal rocks were sent on escaping orbits. Simulations of the trajectories, using the same kinds of software that are used in planning deep-space missions, show that ejected rocks would end up on the Moon, Mars, Venus, and Mercury, sometimes getting there in only a few years, but more often after tens of centuries. So Earth life was transported to Mars. Whether or not it survived there is another story.

Shoemaker deduced this interplanetary swapping of surface

materials in a 1963 paper called "Interplanetary Correlation of Geo-logic Time," written shortly after his PhD work on Meteor Crater. He and his colleagues concluded, in a series of calculations, that meteorites from the early Earth and Mars will be found on the Moon. It is clearly not easy to find them; they are not obvious in the Apollo samples, although certainly a few dust grains are present and possibly some of the granite-like rocks. If large fragments of upper-crustal Earth rocks, or even sedimentary rocks, can be identified and collected from the Moon, they will have been preserved under ideal conditions, serving as time capsules from the dawn of life. "Whether it could ever be recognized is difficult to say," Shoemaker and his colleagues admit.[33] "But the possibility that such material could carry organic hitchhikers, however remote, may present a vexing question to those who are concerned with the origin of life."

How does panspermia work? When life was beginning and the major basins were forming, fragments of Mars were launched that crashed into Earth and vice versa. The pathway from Mars to Earth is preferred on account of Mars's small size and lack of atmosphere, making it easier to blast things off, and the fact that it is exterior to Earth and closer to the bombarding asteroids. But it works in both directions.[34] A large cratering event would produce a spray of small new asteroids, tens to hundreds of meters in diameter, and the conditions for life there would be tough. But having adapted to the extreme environment near the surface of a young planet, *extremophiles* captured on such a ride would be ready for anything! A jolt into space, then a decade or longer encased in tens of meters of rock, then a collision into the atmosphere or the ocean of another planet—a huge jolt—that might seem like a vacation. Other than the radiation of space (from which the rock would provide shielding), it may be no worse than what the surface of the planet would be dishing out in terms of hostile chaos. Tiny organisms encased in porous sedimentary rock might be sheltered from the shock event even as the rock around them explodes, especially if they are landing on a planet with an atmosphere.

Asteroid impacts can extinguish life and can also trigger diasporas on the Earth. The Chicxulub impact wiped out most organisms, but it also would have transported the hardiest life 10,000 kilometers from the impact to populate new regions of the Earth. Cockroaches benefited for sure. More recently, scientists have identified fossilized diatoms high in the Transantarctic Mountains, whose deposition has been proposed[35] as being ejecta from the Eltanin impact 2.1 million years ago, when a kilometer-sized asteroid blasted a 20-kilometer hole in the Southern Ocean. It scoured the seafloor 5 kilometers deep, eroding seamounts and ejecting thousands of cubic kilometers of water, and embedding asteroid fragments into the seafloor to be found by marine geologists examining the drill cores. The hole in the ocean collapsed in about a minute, forming a central peak that would stand taller than Mount Fuji for a poignant moment before collapsing, broadcasting 100-meter tsunamis throughout the hemisphere in an event that has happened maybe three or four times since. Most of the Eltanin ejecta landed back in the ocean, but hundreds of millions of tons of it made it into orbit and fell back on distant shores.

Whether Eltanin ejecta contributed to the Transantarctic diatoms may be impossible to prove. Settling the Mars debate will be no easier. Still, it's a compelling question. Delivery of life to or from Mars requires bigger impacts than Eltanin, but there have been plenty of those. Suppose we find fossilized microbial mats or ancient diatoms on Mars. If they have familiar chemical patterns, then we might be looking at biota delivered to Mars as ejecta from major cratering impacts on the Earth. That is, we might be looking at Earth-originated life, and that would explain its familiarity.

But panspermia is favored to go in the direction from Mars to

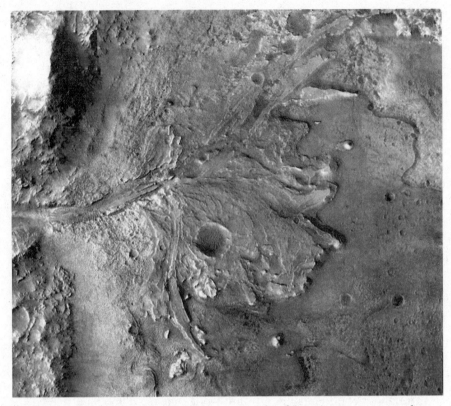

Lake-bed and delta deposits in the west side of Jezero crater, soon-to-be-familiar site of NASA's next Mars rover, which will cache samples for eventual return to Earth. The western crater rim was breached by a wide stream that deposited its bed load of mud and sand into an ancient lake, building up a delta. There has been a combination of large floods and continual flows. Subsequent channels that have gouged the sediments and a 600-meter crater more than 100 meters deep make this a great place to search for fossils. *NASA/JPL/U. Arizona*

Earth. Furthermore, it may be that Mars was ready for life sooner than the Earth, being smaller and cooling off faster from the inferno of radioactive melting and the late stage of giant impacts. If early life existed on Mars in abundance, then it is an incontrovertible conclusion of impact physics and celestial mechanics that it got transported to Earth. If a Martian organism survived in space, and was delivered in sufficient quantities to become established on Earth, then that might be our common ancestor. Controversial? Sure. But this time the hypothesis is testable: the answers might be found in ejecta from the early Earth that were mixed into the oldest regolith of the Moon.

What if we discover something extraordinary and deeply *unfamiliar* on Mars, a fossil life-form or even a living organism that we can prove is completely unrelated to life on Earth—a second genesis? I don't know how we could prove that, but the implication would be that life in the Universe happens everywhere there are Earthlike conditions, within a range of extremes. But if genesis happens everywhere, perhaps it would be just as likely—and a heck of a lot easier—to search for chemical fossils from a "first genesis" right here on Earth, that was replaced by this one, the way *Homo sapiens* nudged out the Neanderthals. There might even have been aboriginal life on Earth that was encroached on by panspermia from Mars; we'd be invasive Martians.

CHAPTER 4

STRANGE PLACES AND SMALL THINGS

T HE CLOSER A PLANETESIMAL IS to the Sun, the faster it orbits. That's Kepler's law. Thus if you're hanging out around 1 AU, where the Earth was born, a planetesimal closer to the Sun will move ahead, and a planetesimal farther away will fall behind. Can these differently paced planetesimals catch up, in order to gravitate together and build a planet? Yes, if their mutual gravity is strong enough. But if *Keplerian shear* is stronger—this tendency for adjacent planetesimals to orbit at different speeds—then the region will not clump into a planet. According to the coagulation equation, derived from balancing shear against gravity, a protoplanetary disk around a sunlike star can give birth to a ten-Earth-mass planet at 1 AU, but a one-Earth-mass planet will be sheared apart faster than it can grow.

Before we can even get around to solving that problem, there is the "meter size barrier." Once a planetesimal grows to be the size of a small car, according to calculations, it will spiral into the Sun in a few dozen years. It's a catch-22: if there's enough dust around to coagulate into planetesimals, then there's enough gas around to drag the planetesimals into the Sun as they form. (It's the same drag force that causes you to slow down in a headwind, and a low-orbiting spacecraft to spiral in.) But if all the planetesimals spiraled into the Sun, why do planets exist? We know of dozens of well-defined planetary systems, and hundreds more waiting for follow-on observations, and about four thousand confirmed planets so far, so the process is common.

The solution seems to be that swarms of planetesimals change the dynamics of the gas and of one another, causing them to cluster and coagulate. A single particle would indeed spiral in, as per the theory, but there's the rub: there is no such thing as a single particle. Instead, millions of particles interact en masse and set up eddies and wakes in the gas, attracting other particles like bicyclists riding in a pack. The aggregates that form out of this, primordial *rubble piles*, are dissipative, embracing incomers the way that a beanbag absorbs energy. Further pebbles impacting into it get stuck. So instead of ruining accretion, gas drag helps the planetesimals pile up and not only survive the headwind but accrete more and more stuff.

This has been called "pebble accretion," and if it is real, then (per the models) you'd expect there to be centimeter- to meter-sized primary components in primitive comets and asteroids. So what count as pebbles? Some have argued that *chondrules* qualify, the sand-sized spheres that are abundant in early meteorites. Chondrules mostly solidified as droplets of melted silicate material, and most of them formed a half million to two million years after the oldest solids, which is actually kind of late to the ball. It seems to me far more likely that chondrules are the by-product of planetesimal accretion,[1] not the beginnings of it. Plus, chondrules are the size of fine beads, too small to be the "pebbles" envisioned by the theory. Others are of the opinion that we have seen pebbles in close-up spacecraft images of comets and asteroids, such as Comet 67P/Churyumov-Gerasimenko,[2] orbited by ESA's Rosetta mission. These images show bumpy textures on the walls of fresh pits and outcrops, resembling piles of meter-sized grapefruit. Asteroid Bennu, the 500-meter diameter target of the OSIRIS-REx mission, also has a texture[3] full of meter-sized "nuggets" that may be weakly cemented together, but until the samples are returned, it's impossible to know just what they are.

Solving the pebble problem would be much easier but for the fact that nature has many ways of making textured cobble. Impact shocks break down rocks, but not into a regular assortment of sizes. Thermal expansion and contraction can granulate a rock, as can volatile outgassing, or decomposition from one phase of ice or mineral to another. Granule-forming processes might be prevalent especially on comets and primitive asteroids that come into the inner solar system, which is where spacecraft have been able to explore them up close so far. This Sun-beaten environment is a complete novelty to comets, and the yoga-ball-sized bumps on the surface of 67P might be a response to solar heating or to vacuum, and might have nothing to do with accretion.

THERE ARE GAPS in our understanding of the accretion of small bodies, and there are gaps in our understanding of the accretion of large ones. Had it not been for missions returning a wide variety of samples from the Moon, we wouldn't have obtained the now-incontrovertible geologic evidence that the Moon was formed in a late-stage giant impact. This turned out to be the key that fit the

A vent hole on Comet 67P/Churyumov-Gerasimenko hints at some of the structure of the materials beneath the surface. The characteristic scale of these bumps is around 3 meters.
ESA/Rosetta/MPS

lock. Accretion began with planetesimals, sure, but it extended all the way up to the merger of father Earth and mother Theia.

There are many pieces of evidence for giant impacts, but one of the strongest is the prediction of a lunar magma ocean. The lunar crust is bimodal in several ways, including its composition, with the highlands made of calcium-aluminum silicates known as feldspars, and the lowlands of the nearside made of basalts and gabbros. If the Moon solidified from a magma ocean—the consequence of a giant impact—then the highlands are well explained as a flotation crust many kilometers thick: piled-up feldspar crystals that floated to the top of the magma ocean as it solidified, the way that ice floats on a lake. Olivine crystals also solidified from the cooling magma, but being denser, they sank to the bottom. If that's what happened, then sandwiched between the solidifying olivine-rich mantle and feldspar-rich crust would be a residual layer that would, according to geochemical experiments, end up with elevated concentrations of potassium (K), rare earth elements (REE), phosphorus (P), uranium, and thorium—elements that are *incompatible* and could not easily find homes in the solidifying crystals. Evidence for such a layer, called KREEP, is seen in many places on the Moon, but almost only on the nearside. The concentration of radioactive elements in the residual layer could have provided the late heating that powered the volcanic flooding of the lowlands, hundreds of millions of years after the rest of the Moon had solidified.

When it was first introduced in the 1970s, the giant impact theory, like plate tectonics, was subject to a lot of skepticism. There was something for everyone to disagree with. Underlying the idea was the powerful implication that instead of forming terrestrial planets planetesimal by planetesimal, or by direct accretion, you first would build dozens of Mercury- to Mars-sized oligarchs, and then, let the games begin! Today this notion of a *late stage* of terrestrial planet formation, oligarchs consuming oligarchs, is the framework for all the major theories of Moon formation. I think it applies deeply to the question of the origin of life, because it might maximize the *diversity* of terrestrial planets—a bizarre menagerie that can't be accounted for by planetesimal accretion alone.

In addition to these ideas of hierarchical merging, of planetesimals cannibalizing into embryos, and further on through increasingly violent collisions into planets, another revelation has been the idea that gas-giant planets, once they form, migrate closer to and farther away from the Sun, like dreamy skaters circling a pond.

Consequently the very structure of the solar system changes, driven by the movements of the outer giants.

It sounds absurd that Jupiter should migrate at all—it is hundreds of times more massive than the Earth and has more angular momentum than the Sun. But it's worse than that. In the "grand tack" model described below, Jupiter wanders in from 3 AU to 1.5 AU and then, with Saturn eventually in tow, it gets pulled back out to 5 AU. Having the giant planets do that can explain a lot of things, especially the compositional and structural gap in the solar system; whether it is true in detail remains to be tested. What is without question, though, is that whatever the terrestrial planets did, they did it under peril of these wandering behemoths.

The cause of giant planet migration sounds unlikely, and represents a victory for populists: it was the gravitational influence of billions of planetesimals that sent Jupiter and Saturn off-kilter. To understand how that might be, let's take a look at the planetesimals themselves—where they have come from and what they have been through, starting with the outer solar system where the giant planets were born (or so we think—we shouldn't be so comfortable believing so).

TRILLIONS OF ICY bodies orbit the distant reaches beyond Neptune. The core population between 30 and 50 AU are called Kuiper Belt Objects, or KBOs; these include Pluto, the ninth-largest body freely orbiting the Sun, and Eris, not as big but the ninth most massive. (In case you are wondering, every one of the major satellites of Saturn, Jupiter, and Neptune are more massive than Pluto.) Most KBOs orbit the Sun in the same general orbital plane as the rest of the planets. Others, notably Eris with its inclination of 44 degrees, and whose whopping eccentricity brings it from 38 to 98 AU, are evidence of a past history that we are still piecing together. After a few roaming plutoids and predicted giants, the outer Kuiper Belt grades into the sparse but much more populous inner Oort Cloud, which extends to tens of thousands of AU, partway to the nearest star system. Somewhere in that outer darkness, hundreds or even thousands of AU away, a cold beast more massive than planet Earth might be lurking, but that's a story for later.

We have never seen any of the comets in the Oort Cloud directly, so the population is theoretical, known only from its members that take steep dives into the inner solar system and then head back out to almost-interstellar space, sometimes blazing phenomenally like Hale-Bopp and Hyakutake. (It's easy to calculate their orbits back out to aphelion.) Cosmochemists would give anything to get a piece of one of these original condensates of the original molecular cloud that predated the Sun. Astronomers observe the glowing ionized gases that are blown from primitive comets by the solar wind, like luminous hair, during perihelion passage.

One of the most interesting KBOs we know of is the rapidly spinning Haumea, with its two known satellites Hi'iaka and Namaka. Haumea is on an orbit similar to Pluto's around the Sun. It spins so fast, 3.9 hours, that it takes the shape of an elongated spheroid almost 2,000 kilometers across. It rotates significantly faster than any solar system body larger than 100 kilometers in diameter, attaining almost the same diameter as Pluto on its longest axis, but its short axis is half as wide. Despite having much less surface area than Pluto or Eris, Haumea is the brightest trans-Neptunian object (TNO), with a surface that's as white as snow. And if that's not sufficient to pique your interest, it's encircled by a debris ring. Moreover, it is dynamically related to a dozen smaller KBOs whose members have the same bright, water-ice-rich surface composition—apparently the smoking gun from a giant impact of some kind.[4]

No general discussion of Pluto is complete without commenting

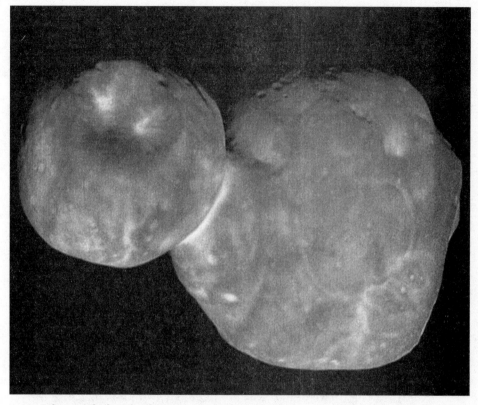

Ultima Thule, also known as 2014 MU69, is the most distant body visited by a spacecraft. This 31-kilometer-long primitive object, over a billion kilometers beyond the orbit of Pluto, is a *contact binary,* formed either in the very earliest accretion or perhaps as a reaccretion following a slow-disruptive collision.
NASA/JHUAPL/SwRI

on what is a planet. In 2006, the International Astronomical Union (IAU) passed a resolution that I paraphrase a little: A planet *is a celestial body that is in orbit around a star, that has sufficient mass for its self-gravity to overcome rigid body forces so that it assumes a round shape, and that has cleared the neighborhood around its orbit. A dwarf planet is as above, but has not cleared the neighborhood around its orbit.* That sounds straightforward, but is it? First we should amend it to exclude stars that are orbiting stars; these aren't planets. At the other extreme, we have to exclude a blob of water floating inside of a spaceship, so as to not call that a dwarf planet. Fair enough, but what about Pluto? Its gravity has made it nearly round, so it's good on that front. It also has very few impact craters, an expression of its active geological evolution.

So there's the first problem. You'll notice the IAU definition has nothing to say about geology, and from the geological perspective Pluto is a planet.[5] The Astronomical Union definition also misses the mark in calling non-planets "dwarf planets," because that distinction has nothing to do with a planet's size. Pluto is called a dwarf planet because it is dynamically shepherded by Neptune. If the IAU definition holds, then when we discover an Earth-mass planet orbiting in the habitable zone of another sun, with liquid water on its surface, that is under the gravitational influence of a super Jupiter, then the rule will be to call it a dwarf planet. That would be silly.

IN THE LATE 1980s, Pluto was on astronomers' minds, because a series of mutual eclipses by Pluto and its satellite Charon had just ended. (No images existed; they were still just dots of light.) Their orbits and masses were now precisely known, and from the eclipses, their diameters. The bulk density of Pluto could be computed: 1.9 grams per cubic centimeter, halfway between ice and rock, slightly less dense than Triton, slightly bigger than Pluto, that was visited by the Voyager 2 flyby of the Neptunian satellites in 1989. Telescopic data from these eclipses were used to derive the first crude geologic maps of Pluto and Charon; looking back at those images gives us a sense of what the first Earth-analogue exoplanet images are going to look like. By applying least squares, astronomers derived a color map of sorts[6] that proved without question that Pluto has a variegated geology indicative of a long and possibly ongoing surface evolution.

Tombaugh Regio, named after Pluto's discoverer, and the adjacent Sputnik Planum make a bright heart on the surface of Pluto. Charon in the back left is darker, acquiring organic carbonaceous molecules lost from the bigger planet. Composite of New Horizons multispectral imager observations. *NASA/JHUAPL/SwRI*

Herewith a dynamical puzzle. Pluto's orbit being so eccentric, it crosses inside the orbit of Neptune for one-twelfth of its year, which is 248 Earth years. You might think that planets on crossing orbits would collide, eventually, but in this case Neptune and Pluto are synched up in a 3:2 resonance, so Pluto crosses in and out only when Neptune is far ahead or far behind. Why would a planet that is 2,300 kilometers in diameter, orbited by a moon half that size, be found on a highly inclined and eccentric orbit around the Sun, in a reliable orbital resonance with Neptune? Pluto-sized planets are able to form only in the midplane of a protoplanetary disk, because that's where all the matter is, so Pluto must have been somehow skewed out of its original orbit after its formation.

The first idea was to attribute it to a gravitational "slingshot," when a massive KBO had a close encounter with Neptune. Another idea was that Pluto escaped from orbit around Neptune and is a brother of Triton, which is about the same mass and orbits Neptune in the *retrograde* direction, counter to the planet's rotation. Definitely something weird must have happened[7] to explain Triton's backward orbit. But each of these scenarios—an escaped satellite, or a Neptune-grazing body—is dynamically impossible. Getting from an orbit that encounters Neptune to one that is precluded from encountering Neptune is like doing a bank shot from the wrong end of the table. To use the language of dynamics, it's on the other side of the *separatrix*.

In the early 1990s the American astrophysicist Renu Malhotra came up with an idea that would connect the excited orbit of Pluto, an old problem, with the new concept that was emerging, of giant planet migration by the scattering of planetesimals. According to her theory, Neptune accreted in the midst of a swarm of icy bodies orbiting the Sun. Every once in a while a close-flying small planetesimal made a slingshot orbit around Neptune and was scattered away. There were many billions of these encounters, and each one tugged on Neptune a little bit. Because Neptune is so massive, these nudges were small, but they added up to a net force because the nudges were asymmetric, most of the small bodies coming from outside Neptune's orbit. (The ones closer to the Sun had been consumed or scattered in the formation of the other giant planets.)

So there was a small imbalance, a net force going out instead of in. For reasonable starting populations of planetesimals, calculations showed that Neptune could migrate continuously from a starting location around 20 AU, outward by 7 AU or more until the planetesimals were scattered away or were too sparse to matter.

Pluto, born outside of Neptune's starting orbit, was in the way. So were a lot of other major bodies. But before Neptune could catch up, they got caught in a situation where Neptune orbited the Sun three times for every two times Pluto did, which resulted in a powerful gravitational coupling in which Neptune exerted an unbalanced influence on Pluto. As Neptune's orbit expanded, it pushed on Pluto, expanding its orbit and increasing its eccentricity and inclination in the way that a kid in a swing starts to wobble left and right if you push him as high as he wants you to.

Many of the largest original KBOs did not get so lucky but were accreted by Neptune or by another planet, or were ejected from the solar system altogether. Still, Pluto is not alone in pulling off this artful dodge; more than a dozen "plutinos," smaller than Pluto, have been trapped into a resonance with Neptune, making an agreement with the twenty-Earth-mass beast that shelters them from further chaos, never to get in its way.

The success of this model implies that Jupiter and Saturn also migrated after they formed, by these same kinds of planetesimal and cometesimal interactions. Born in the thick of the solar system, they didn't always migrate out, like Neptune, but sometimes in. The first application of this idea, that Jupiter and Saturn underwent significant migration, is named the *Nice model* after the observatory in France where the idea came together. It was recognized that like the orbital periods of Pluto and Neptune, only more so, the orbits of Jupiter and Saturn might beat together during this migration. Depending on how you start things (the starting locations of the planets; the mass distribution of planetesimals, playing

God, as modelers do), the two largest planets migrate through an extremely powerful and lopsided 2:1 resonance, Jupiter orbiting twice for every orbit of Saturn. They would intermittently line up on one side of the Sun or the other, these two biggest masses in the system, turning the planetary system into a gravitational motor with the camshaft out of balance. (If Jupiter represents the motor, then Saturn's original system of moons, in the scenario below, could be the nuts and bolts that vibrate loose.)

The Nice model is good at explaining the structure of the trans-Neptunian population, because it scatters a disk of objects onto inclined eccentric orbits, where many of them are, and leaves a "classical" disk around the midplane. It can also throw light upon the origins of Uranus and Neptune, because if the ice giants originated where they are today, it is difficult to explain how they formed. The problem is that the material from which they accreted would have been so spread out, at 20 and 30 AU, and orbiting so slowly, every 100 years or longer, that Neptune would take 10 billion years to form. In the original Nice model, the problem is solved because Neptune and Uranus were formed in between Saturn and Jupiter, and were ejected when the 2:1 resonance imbalanced the system. But while the Nice model can explain why Neptune exists, it disrupts the elegant story for Pluto just told. In the convoluted field of planetary formation dynamics, there are many evolving models to keep track of.

The most famous prediction of the Nice model is probably wrong.[8] One of the first observations from laboratory analysis of the Apollo samples is that they showed a commonality of ages, with many of the impact melts dating to around 3.9 billion years ago. This was thought to be the record of an enormous spike in the impact rate, that became known as the Late Heavy Bombardment. Given the timing, this LHB would have happened hundreds of millions of years after planet formation had quieted down—and coincidentally, right when life began flourishing according to the geological record of the Earth, making it a profoundly important theory. Imbrium and five or six comparable basins may have formed on the Moon, with an even more violent pummeling of the Earth, and an equally intense cratering on Mars and other planets.[9]

Today there is a lot of debate about whether there really is a cluster of ages at 3.9 billion years. For one thing, Apollo returned samples from only six nearside locations; when we include the growing collection of lunar meteorites, there is no single spike in ages[10] but rather a few random spikes from around 2.7 billion to 4.2 billion years ago that can be explained by stochastic events like the disruption of a major asteroid producing a swarm. For another, if you look at each Apollo sample in finer and finer detail, using modern "nano" instruments, the ages obtained are more diverse. Suppose that you have a thousand red, green, and blue socks randomly mixed in a pile. If you pull out individual socks (small samples), you would conclude, "Wow, these socks [ages] are really diverse!" If you pull out piles of a hundred socks each time—the equivalent of the earlier measurements—there would be about one-third of each color in each pile, and you would conclude, "Wow, these clumps of socks look more or less the same!"

If there was an LHB, then the Nice model offers an explanation. The idea is that the Jupiter-Saturn resonance waited to happen for about 650 million years after planet formation. Early migration got them close to the danger zone, but it took another half billion years to go the last fraction of an AU and fall into the dreadful 2:1 resonance. And then, with this grand alignment switched on, the solar system—which seemed to have finished forming—exploded with a last round of dynamical activity and planetary rearrangement. The solar system became a hail of projectiles until Jupiter and Saturn exited the resonance, like two gigantic ships leaving a storm of their own making, and things got back to normal.

A fundamental problem with the Nice model[11] was identified as soon as it was published. If the giant planets migrated in this manner, they would migrate into resonances with the terrestrial planets as well. If Jupiter ended up in an orbital resonance with Earth, for example, this might have pumped up the eccentricity of Earth's orbit to values ten times greater than it is today. Venus is even harder to explain; its eccentricity is today nearly zero, but its orbit should have ended up strongly excited. Poor Mars would have been dragged along like a kid brother at an amusement park, and even become Earth-crossing in some calculations. During perihelion (closest to the Sun, *helios*), it would then receive twice as much heat for a couple of super-summer months of each year. Then it would go off into the deep freeze, passing through the Main Belt for a couple of years. Nobody to my knowledge has modeled the climate of a Mars gone astray, but these extreme temperature swings, like a casserole that gets put in the freezer, then thawed, then frozen again, might trigger epochs of cataclysmic landform evolution consistent with the mysterious mega-flooding of ancient Mars.

Whatever you make of that implication for Mars, you can't have a model that ends up with Earth and Venus on excited orbits. Either it didn't happen that way or something must have damped them. And there are other problems. The LHB would, according to the Nice model, be solar-system-wide, but then the swarms of interloping small bodies would have wiped out the inner Saturnian satellites, destroying Enceladus and Mimas many times over.

Despite these substantial problems, the Nice model has developed a certain resilience by endlessly transforming. In 2010 it was proposed that instead of continuous migration of giant planets, their migration is impulsive, mediated by three or four Neptune-like planets, themselves dragged around by planetesimals, of which two have survived. "Jumping Jupiters," they call this model. Every time an ice giant migrates close to Jupiter or Saturn, the mutual gravitation is such that planets react by "jumping" (over the course of a few thousand years) to different orbits, like giving the radio dial a good twist. This would allow the desirable features of the original Nice model without them ever landing on a resonant frequency with Earth and Venus.

If there wasn't an LHB, then instead of the Jupiter-Saturn resonance having to happen around 3.9 billion years ago, this dynamical explosion of the solar system could have happened immediately after planet formation, shuffling around the planets in a final churn of wandering orbits and resonant companions, an extended origins story. The model would still be able to eject Uranus and Neptune, scatter KBOs to the Kuiper Belt, capture the irregular satellites of Jupiter—but would not happen late. Still, when all's said and done, it may be Neptune's connection with Pluto that ultimately anchors any theory for the outer solar system, because its gradual outward migration is required to sweep up Pluto and the plutinos. All of these stories have yet to hang together.

MARS BOUNDS THE giant planets on the inside and is therefore a gateway to the outer solar system. It has two very weird moons, discovered in 1877 by Asaph Hall using the newly commissioned twenty-six-inch refracting telescope at the U.S. Naval Observatory. He named them Phobos and Deimos after the loyal and bloody sons of Ares, god of war. And here begins one of the stranger stories in planetary science. A century and a half earlier, long before any telescope could have seen them, Jonathan Swift wrote in *Gulliver's Travels* about two small moons of Mars that had been discovered by the astronomers of the flying island of Laputa, "two lesser stars, or satellites, which revolve about Mars," that he went on to describe. Swift could have had no idea, but in *Gulliver's Travels* he actually came up with two satellites close to their actual positions.[12]

The inner one reported by Gulliver orbits at 3 Mars radii every ten hours. Phobos actually orbits at 2.8 Mars radii every eight hours. The outer one in the story orbits at 5 Mars radii, compared to 6.9 for Deimos.

One idea for this coincidence is that Swift got the idea from his interest in the writings of Kepler, who had passed away a century before. Kepler paid significant attention to the work of Galileo, who in 1610 was making discoveries every clear night. Scientists were as competitive then as they are today, and faced similar delays in the publication of their results, and getting scooped, so a form of crypto-publication became popular, in which one sent around scrambled anagrams as a means of reporting results while still working on the measurements and analyzing data. That way, if and when their discovery was scooped, they could support the claim to being first by descrambling the anagram.[13]

That year, 1610, Galileo sent the following jumble of letters to Kepler and other colleagues: *smaismrmilmepoetaleumibunenugttauiras*. Unjumble as intended, and it reads *"altissimum planetam tergeminum observavi,"* or "I have discovered the highest planet to be triple-bodied." He had discovered "knobs" around Saturn, which later observations would show to be rings. Kepler was perhaps expecting news about Mars. Profoundly influenced by numerology, he believed that if Earth has one satellite and Jupiter has four, then Mars, in between, must have two.[14] So he decoded Galileo's anagram optimistically, off by one letter and in stilted Latin: *salve umbistineum geminatum Martia proles,* or roughly, "hail, twin knobs, children of Mars." So Jonathan Swift—a great fan of Kepler, not just his science but his posthumously published science fiction novella *Somnium Astronomicum* mentioned in chapter 1—might have known of Kepler's ideas and lucked out on the orbital distances.

Phobos and Deimos were the first small bodies imaged by a spacecraft. By "small bodies," I mean objects that are irregular in shape, without enough gravity to force themselves into a spheroidal figure. Phobos and Deimos are not asteroids per se, although not long ago the mainstream idea was that they were interlopers captured from the Main Belt. They sure look like asteroids, these dark reddish cratered clods. But capture around a planet is extremely unlikely. It's like throwing a tennis ball into the open window of a moving car several streets away, but a lot harder. If the encounter was too fast, they would not be captured; if too slow, they would impact the planet. And even if capture could work, asteroids come at Mars from all directions; it would be a stunning coincidence for either of them, let alone both of them, to end up in the equatorial plane of Mars and have circular orbits.

Scientists have proposed since the 1980s that a giant impact created the Borealis Basin of Mars, the lowlands of the northern hemisphere that defines another geological dichotomy. According to computer models, the impact formation of a basin that large would have produced a debris disk with more than enough material to make Phobos and Deimos, material that would have aligned with the equator of the planet. But this introduces a new problem: producing the Borealis Basin would eject at least a thousand times more material into orbit, enough to make a satellite that is hundreds of kilometers in diameter, proportional to the mass of the Moon around Earth. A much smaller impact event doesn't make a satellite-forming disk at all—the ejecta either falls back down or escapes—so it seems to be all or nothing, either a major satellite of Mars or none. We'll revisit this question later, but for now let's savor it as one of those delightful scientific problems where a seemingly tiny thing—an oddity like Phobos—changes how you think about a bigger thing like Mars, then changes how you think about planet formation itself—tails that wag the dog.

Why *is* Mars so small, when the expectation is that a fat disk of material would have originally extended between Venus and

Saturn? If the disk extended uniformly, without a gap, then Mars should have grown to become five or ten times more massive and much more water rich—more Earthlike—instead of small and dry. Enter another, even weirder theory based on giant planet migration. The grand tack theory[15] has Jupiter born at 3 AU, not 5 AU, where it is today. Instead an ice core, it would have a rock core. From there, it would migrate in to 1.5 AU, where Mars is today, wiping out almost everything that was there. Then Saturn would be born and would itself start moving in. Saturn would then get caught up in a 3:2 resonance with Jupiter, where it would then get stuck, in the manner that Pluto and Neptune get stuck but a more equivalent mass ratio. This would create a powerful torque on the orbits of Jupiter and Saturn, dragging them *out*, putting both on track to where they are today.

So taking a gigantic step back, where do we stand? Giant planet migration occurred, and it changed the original architecture of the solar system. Scientists have come up with models that can explain certain things like the small mass of Mars, or the existence of a scattered population of KBOs, or the resonant orbits of Neptune and Pluto, or the Late Heavy Bombardment of the Moon if that occurred.[16] But as for how and when it happened and in what order, and what changes it caused, there is great uncertainty. Add to that the surmounting evidence that our solar system is unusual, and the knobs can come off the dials controlling the models.

DESPITE THIS AMBIGUITY about where it was born and how far it has traveled, Jupiter may be an archetype for how satellite formation plays out around gas giant planets throughout the galaxy. The modern view is that the Galilean satellites were created toward the end of Jupiter's formation, long before the orbital rearrangements just described, out of a massive sub-nebula of ice and dust around the planet. The earliest of these satellites didn't make it, or rather, did make it, because once they were formed, they were dragged into Jupiter by their gravitational interactions with the disk.[17] (If massive enough, a moon would set up density waves in the gas disk, creating an asymmetrical force similar to the effect of tides.)

These massive satellites, thousands of kilometers diameter, spiraled in to Jupiter one after another, part of the late accretion of the planet. But Jupiter's metallic hydrogen core, spinning rapidly and interacting with the solar field, generated a potent dynamo,[18] producing a magnetic field that cleared away a "doughnut hole" in the gas and dust closest to the planet. Once this gap opened up around Jupiter (goes the theory), the disk-driven inward migration of the newly formed moons stopped, and when the next moon was created out of the dust and ice, it parked at the edge of the gap. This was Io.

Although it started as an ice-rich world, Io was now trapped close to the furnace-like heat from the radiating newborn Jupiter. The next icy body to spiral in was Europa, whose own migration ended when it was captured into a 2:1 orbital resonance with Io, like a needle finding its way into the groove of a record. Because it parked farther from Jupiter and came in later, much of Europa's water survived. Then came Ganymede, named after the young shepherd that Zeus abducted to become his cupbearer (the Galilean satellites are named for the mortal lovers of Zeus); its migration ended in a 2:1 resonance with Europa.

This stacked triplet forms what is known as a *Laplace resonance,* named after the French polymath Pierre-Simon Laplace, who proved the powerful stability of such a trio. According to his theory, the fourth Galilean satellite (lovely Callisto) would then be the last to spiral in. But by this time the nebula was gone, so Callisto's migration ended before she could catch up with Ganymede.

If most giant planets in the Universe form the way Jupiter did, we could expect that they would commonly produce systems of satellites that end up in resonant chains. This is important because it gives the satellites a long-lived source of internal tidal heating, which could sustain life.

Satellites in a Laplace chain end up on eccentric orbits, not circular ones. They experience high tides and low tides as they are flexed by Jupiter during every orbit around the planet. This results in tidal dissipation, which would cause their orbits to expand, except that one satellite can't migrate without the others in the resonance. The result is like a well-wound clock that will keep moving stably for tens of billions of years, longer than the age of the solar system. The frictional heating by tides is substantial, and isn't theoretical—just look at Io, a satellite the size of the Moon, only 5 planetary radii from Jupiter, covered with volcanoes, the most geologically active body in the solar system.

Tidal heating drops off sharply with distance from Jupiter: the tidal forces are smaller and each orbit takes longer, so there are fewer tidal oscillations. The tidal heating in Europa is therefore modest compared to Io's heat flow. Together with the radioactive heating, it is sufficient to allow a liquid water ocean to exist beneath an icy carapace. (Because water floats when it freezes, it forms an insulating lid that keeps in the heat.) Ganymede is also in the Laplace resonance, and although it is heated less intensely, it also has a liquid water ocean supported by stronger radioactive heating. In fact, its spectacular record of geological activity may date to when Ganymede first got locked into the Laplace resonance and responded to the shock of tidal heating by erupting with floods and extrusions of viscous brine.[19]

Just as the Moon migrates away from the Earth, because of tides and the Earth's rotation, the Galilean satellites migrate gradually away from Jupiter. They do it in lockstep, with none of them collid-

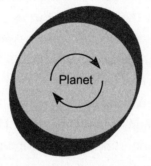

A tidal bulge raised on a planet is shown rotating ahead of the sublunar point. This causes an unbalanced gravitational force that applies a torque to the moon, causing it to spiral out (in this case). The moon has its own bulge, but when it is tidally locked (which happens fast), it would be frozen in place, pointing at the planet as shown. But when satellite orbits are eccentric, the tidal bulge (exaggerated in the figure) wobbles to and fro. This repeated deformation causes friction that is responsible for the hyperactive volcanism of Io and drives geology on other satellites. Tidal heating was intense enough inside the early Moon to have kept it molten. Since then (due to the bulge it raises on Earth) it has migrated out to a dozen page-widths away (at this scale), so that now the heating is minor.

ing or getting into other dynamical trouble. But their orbits always remain eccentric, a situation that is predicted to outlast the Sun; at 5 AU they might not even be greatly disturbed by its red giant phase. From the perspective of the origin and sustenance of life, the geologic evidence for a long, even burn of orbital-tidal energy means that wherever gas giant planets exist beyond the ice line, they might have moons sustaining ocean environments lasting billions of years.

WHAT IF SOLID H_2O sank? These outer solar system oceans, now open to space, would radiate away all their heat in ten thousand years, freezing from the bottom up and exposing anything at the surface to brutal radiation. They would become lifeless, desiccated rinds. Instead the ice crusts float, forming a strong, stable, highly insulating shield. Extremely cold ice is actually as hard as granite, and any breach—for instance, an impact crater—rapidly freezes over and anneals. At the surface of Europa, it is a stunning 200° below zero. A few kilometers below this cryocrust begins an ocean equal in volume to all the seas on Earth.

At first the evidence for Europa's ocean was circumstantial: Europa's surface is made of ice, and it has the bulk density of a body mantled by hundreds of kilometers of H_2O. It shows geologic evidence for liquid water interacting with an ice crust at and just below the surface, in the recent past, and comets appear to have punched into what appears to have been liquid water. It's also predicted to be strongly heated by Jupiter's tides, both in theory and by analogy with nearby Io, which is covered in volcanoes. Even if all this evidence adds up, it points to the past, an ocean that *once* existed on Europa. But what about today?

In addition to taking images, the Galileo mission acquired all kinds of other data. The American planetary physicist Margaret Kivelson showed that the measured magnetic field matched the prediction for a moon with an electrically conductive ocean— namely a brine layer 100 kilometers thick, of a certain salinity[20]— orbiting within Jupiter's powerful magnetic field, twenty thousand times stronger than the Earth's.[21] This direct measurement of a conducting layer proved the existence of the first global ocean beyond the Earth.

Europa has a tenth of the gravity of Earth, and its ocean is ten times deeper, so the pressure at depth is comparable to the oceans of the Earth. And perhaps surprisingly, at the base of its ice shell,

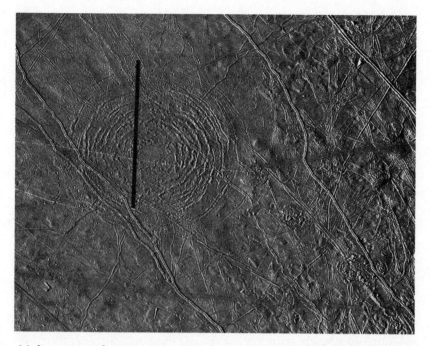

Multi-ring scar from a comet impact, a resurfaced splotch on Europa called Tyre Macula, 50 kilometers diameter at the outer edge. The black bar is a data gap. Tyre is tens of millions of years old, formed by a 5-kilometer comet striking the ice shell, somewhat like throwing a rock into an ice-covered pond.
NASA/JPL

its ocean is not much colder than at the base of the ice shelves on Earth, a couple of degrees below zero. Any colder and it would be a solid. The salinity of its water, estimated from its conductivity, could be brackish (1 gram of salt per liter) or hypersaline (100 grams); seawater on Earth has 30 grams of salt per liter, so these could be familiar conditions. The ocean's geophysical characteristics (how it circulates, how it interacts with the fresh-ice shell) also depend on salinity, which determines the melting temperature and density.

Tidal heating of Europa is thought to happen mostly in the base of the ice shell, where tidal flexure generates most of its friction, with radioactive heat coming from the rocks of the mantle. The ocean receives no heat or light from above, and except for possible triboluminescence (when solids that are grinding together make light) and bioluminescence (when creatures make light), the ocean of an icy moon is pitch-dark forever. Communication with the surface appears sealed off. But there's plenty of evidence that water has breached the surface of Europa in the past—for example, the so-called chaos regions, ubiquitous areas that look like a frozen backyard marsh in the winter after the dogs have had at it.

Chaos ruptures on Europa might be infrequent; the Galileo spacecraft was able to image only a couple of them in detail, and we haven't been back. But more recent telescopic observations suggest that water could be making its way through local pull-apart fractures. The detection of trace amounts of oxygen in plumes around Europa indicates small but recurring events—a recent or ongoing water-rich eruption is the best explanation. Europa is being monitored for further evidence of outbursts, which could lead to new ideas for missions.

The surface of Europa is a horrible place for life. Jupiter's gigantic magnetic field traps charged particles from the Sun, funneling them in to create a blazing bath of concentrated radiation intense enough to break apart molecules and ionize atoms. It will kill anything that ventures to the surface and is intense enough to fry even a hardened robotic explorer. That is why a Europa lander has not been attempted; half of its mass would have to be radiation shielding. Even so, if and when we do land on Europa, we have to be careful not to infect the planet with native Earth bugs. Just landing is probably okay—infecting the near surface is impossible due to the sterilizing radiation. You can probably land, do all the seismology and geochemistry you want, and any organism (virus or bacterium) that happened to hitchhike would have its DNA torn to shreds. But many advocate going much further—penetrating through the crust and heading down into the brackish ocean[22]—and for that we have to be patient.

Russian scientists have drilled into Lake Vostok, a lake on Earth that is 4 kilometers below the Antarctic ice sheet. If we want to search for a second genesis in the solar system, this would be an excellent goal, perhaps sealed off from the rest of life on Earth for millions of years. Sadly, the urge to explore within our lifetime has overwhelmed the sense of scientific caution, and we are on the verge of contaminating these absolutely unique biomes because humans are impatient. It is like the archaeologists of the 1800s plundering the tombs of Egypt, except worse. A few stray microbes could very well thrive there (in Lake Vostok, or in Europa), and we could destroy the only truly alien ecosystem that we could ever hope to study in a clumsy attempt to discover it.

FOR CREATURES THAT may be teeming in Europa's inky depths, hell is *up*, above the cryosphere. Heaven is *down*, under the lid, and near the black smokers and seamounts at the rock-water interface. But every once in a while, this "octopus's garden" would get all mixed up. How often, we can guess from Europa's sparse craters. Astronomers have a good handle on the number of comets coming into the Jupiter region per million years. We can then count how many craters Europa has per million square kilometers. Combined, these indicate a surface age of around seventy million years, with no region a whole lot older or younger than another.[23]

What happened? It may have experienced a global geologic upheaval at that time, of the sort that is also hypothesized for Venus—perhaps a major comet impact or collision by a lost satellite of Jupiter. Another idea is that the ocean could become thermally or compositionally stratified, growing gravitationally unstable in time, so that every once in a while the whole ocean overturns and upends the floating crust along with it. It's conceivable that life inside Europa feasts episodically upon a surface bounty. The surface is implanted by comets and dust, and these materials are made into new compounds by solar radiation and cosmic rays. Maybe deep inside, a dormant biosphere is waiting patiently for the next time that this batch of ripening crust is upended.

The oldest large craters on Europa—tens of millions of years old—look like they punched through to the liquid water ocean, leaving little if any final topography. Since then the ice shell has apparently been thickening, to the point that the most recent 30-kilometer crater, Pwyll, looks like it formed in solid ice at least 10 kilometers thick. As time goes by the thickening of the ice lid happens more slowly, since the heat stops getting out and it happens irregularly; today it appears Europa's ice crust varies from a couple of kilometers thick in places, to only a few hundred meters in others, to 10 or more kilometers everywhere else.

Explorers are most interested in the thin-ice places, wherever ice might have been breached most recently. NASA's upcoming Clipper mission will make a series of flybys of Europa in the late 2020s, to get combined radar and magnetic detections of the electrically conducting brine beneath the radar-transparent ice crust. Shallow areas will stand out as potential future landing sites.

Ganymede, three times the mass of Europa, has no surface expression of a shallow ocean; the existence of water is deduced from theory, and from the magnetic field measurements indicating a conducting brine. From the deformation of craters and opening of cracks, Ganymede's crust appears to be solid ice at least 50 to 100 kilometers deep—is that too deep to care about? That's several times deeper than the limits of the biosphere of Earth. To breach the ocean inside Ganymede would require the impact by a massive body so uncommon that such a collision might not happen again until the Sun is old. If there are inhabitants living on the base of the Ganymede ice crust, they might never discover that up is down and down is out, even if they outlive all the other planets.

THE GALILEAN SATELLITES, once they formed around Jupiter, are such massive bodies, and in such a stable configuration, that no event in the solar system, short of destroying them, is able to unlock them from their Laplace resonance. They have remained in synch around Jupiter since their formation, regardless of where Jupiter has migrated, and how. Their movement and the tidal heating of their oceans shall outlast the death of the Sun.

If we are satisfied with the explanation for Jupiter having a Laplace chain of massive satellites, then why does Saturn have only one large moon, Titan, and a smattering of siblings? One explanation is that Saturn was too small to generate the intense magnetodynamo needed to clear away a gap in the inner disk. Without a gap, the moons kept spiraling in. Titan—so goes the theory—would be the last one caught in the act when the gas went away. But there's not only Titan but also a set of middle-sized moons (MSMs) to explain, and these are a class of objects that Jupiter does not possess.

The most significant MSM in terms of figuring out the origin of the Saturn system is probably Iapetus, almost half the size of the Moon, that orbits three times farther out than Titan, and on a substantially inclined orbit. That all has to be explained. The density of Iapetus is slightly greater than that of ice, and some of its surface is blindingly white while some is black as pitch. To give a sense of its strangeness, after the highest-resolution Cassini images came down, mission scientists ended up scratching their heads via emails, trying to come up with something to say to the press. Several argued that the black features, made of organic material darker than charcoal, were pasted onto snow-white terrain.

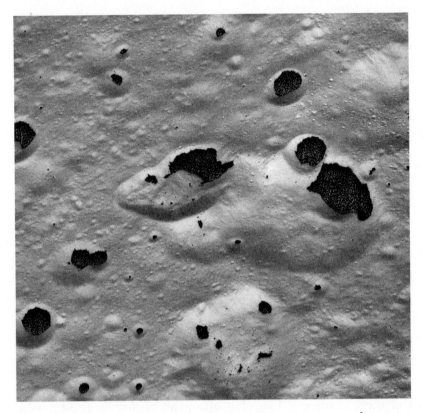

A 35-kilometer patch of a transition zone on Iapetus, a massive distant icy moon of Saturn. Black on white or white on black?
NASA/JPL

Others argued for the opposite, white on black, and each camp had its theory to support it.

Each of Saturn's middle-sized moons has its own bizarre geology, and there is nothing systematic about *how* they are so different, which is also deeply puzzling. Iapetus is just the beginning of this rabbit hole of weirdness, one of the "seven dwarves," although in this case named after Greek Giants and Titans: Mimas, Enceladus, Dione, Rhea, Titan, Hyperion, Iapetus, and Phoebe.[24] (The last one, on an odd orbit and smaller than the rest, is probably a captured Kuiper Belt Object.) Saturn's moons are of such an ex-

traordinary variety that they represent, at least to me, the best laboratory we have so far for studying the genesis of planetary diversity.

The Cassini mission spent thirteen years orbiting Saturn, and
this gave a precise baseline that supported astronomical observations going back hundreds of years, the timings of satellites wink-
ing out behind Saturn. This has led to good estimates of the tidal
evolution of satellite orbits. We have *seen* Saturn force its satellites
around by tides. To move them so fast, the interior of Saturn must
be a thousand times more dissipative than Jupiter; that is, it must
have a lot more internal friction, so that a thousand times more energy gets transferred to the orbits of its satellites, and migration
happens fast. Although we don't understand the reason for this
difference, Saturn has a substantially different interior than Jupiter, and it generates a much smaller magnetic field. After all, as
Newton first proved so long ago, its bulk density is half the density
of Jupiter, 0.7 grams per cubic centimeter, less than water.

Here's a quick geologic survey of the major moons of Saturn from
the inside out; see if it makes any sense. First is the geologically
dead little planet Mimas, 400 kilometers in diameter, the density
of water ice. Next is one of the most geologically active worlds in
the solar system, Enceladus, 500 kilometers in diameter, with a
density of half rock, half ice, that erupts constantly with geysers
of ammonia-rich water that come from a global ocean. After that
is Tethys, which like Mimas is made primarily of solid water ice,
but is twenty times as massive. Next are the largest MSMs, Dione
and Rhea, both approximately half rock, half ice, both strange and
complex yet nothing like each other. Next is our familiar friend
gigantic Titan, 5,000 kilometers in diameter, substantially larger
than the planet Mercury, ten times the mass of Pluto, and an excellent analogue of Earth in many respects.

Titan is about the same size as Ganymede and Callisto, the largest icy moons of Jupiter, but in stark contrast it is an incredibly

active planet with a dense nitrogen atmosphere and icy continents and methane seas. It has great lakes hundreds of kilometers across, filled with hydrocarbons, on a crust of rigid water ice that is tens of kilometers thick. Deeper still, heated internally by tidal friction and radioactive decay, and subject to greater pressure, Titan's ice shell is believed to transition to liquid, like on Europa and Ganymede. In other words, Titan is thought to have two kinds of oceans: hydrocarbon seas on top, with their interconnected swamps and drainages and aquifers, and then a rigid ice crust, and then a liquid ocean.

Beyond Titan, and captured with it in a 4:3 orbital resonance around Saturn, is the irregular tumbling body Hyperion, much less dense than ice, a sign that it has substantial porosity. Finally— the fly in the ointment of many a theory—is Iapetus, orbiting at the same distance from Saturn as the Moon scaled to the Earth, 60 planet radii, and with high orbital inclination. Beyond Iapetus, there's the irregular-shaped Phoebe, which appears to be a captured rogue object from the Kuiper Belt, and beyond that, there is a host of dozens of irregular moons, which are also likely captured.

The major satellites of Saturn lack a particular trend of anything, whether that be with distance from Saturn or diameter or composition. It's as though God decided to make one random MSM and put it *here,* and then make another one, maybe this time with a thicker ice crust and less rock, and put it *there.* But however they formed, their placement is actually intricately connected, with Titan driving Hyperion in a 4:3 resonance, and Dione exciting Enceladus in a 2:1 resonance, and Tethys exciting Mimas in a 2:1 resonance, and so on. Strange to say, but Tethys and Dione have their own pairs of co-orbital moons (small bodies, one more massive than the other) that share their orbits around Saturn. Their co-orbitals are stable for now; they might have been debris captured after a collision, which is an idea we'll get back to when considering our own Moon.

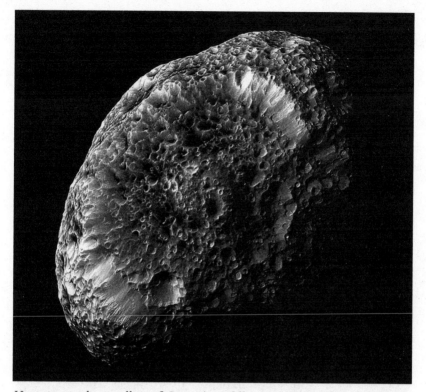

Hyperion is the smallest of Saturn's middle-sized moons, 270 kilometers average diameter, and has a bulk density just over half that of ice, the same as a comet nucleus. It is highly porous. Features down to about 1 kilometer can be resolved. It orbits in a 4:3 resonance with Titan. The flat-walled ridges face toward what could be an old impact blast zone or a global structural collapse. Small depressions are probably mostly impact craters that seem to have grown into suncups, formed when small pits expand as the ice sublimates away. There is much yet to discover about Hyperion and its relationship with Titan.
NASA/JPL/Space Science Institute

The Saturn system offers some potential analogies to the solar system at large. Let Mimas and Enceladus represent Mercury; then Tethys, Dione, and Rhea—the largest MSMs—can stand in for Venus, Earth, and Mars, while Titan is like Jupiter and Saturn, and Iapetus is like Uranus and Neptune. To be sure, it's a stretch. The actual planets are much bigger, for one thing. For another,

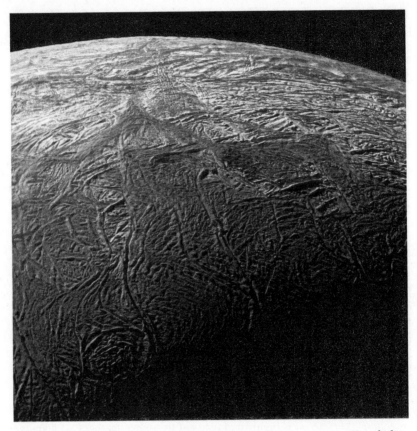

South polar region of Saturn's 500-kilometer diameter moon Enceladus, taken by the Cassini mission in 2009. The "tiger stripes," including Baghdad Sulcus, the source of the amazing geysers, are in the foreground. These show up as glowing stripes in the infrared camera.
NAPA/JPL/SSI

they have only hints of the kinds of interconnected orbital resonances that the MSMs experience so profoundly. For another, Saturn interacts tidally with its moons, heating them and changing their orbits, something that doesn't happen to planets that orbit tens of millions of miles from their star.

Enceladus, only a tenth of a percent of the mass of the Moon, is an outrageously active little satellite. Its heating is produced, we think, because it is forced into an eccentric orbit by ten-times-more-massive Dione, causing tidal friction that grinds away at the base of its ice shell, generating an estimated twenty gigawatts of heat[25] in the "tiger stripes" region around its south pole. But here's the problem: if tidal friction can explain the spectacular geologic activity of Enceladus, then by applying the same calculation, Mimas should be experiencing even greater tidal heating, as it is forced into an eccentric orbit by its 2:1 resonance with Tethys. Mimas should be a cryovolcanic marvel, but instead looks for all the world like a four-billion-year-old spheroid battered by ancient craters, with its cyclopean crater Herschel making it look like the Death Star in the *Star Wars* movies. Another lock without a recognizable key.

Why are there a half dozen middle-sized moons around Saturn? Why does Jupiter have no middle-sized moons? Maybe the two systems were built in different factories by different rules, Jupiter like a station wagon and Saturn like a motorcycle. But they have one thing in common: the mass of Titan, compared to Saturn, is equal to the mass of all four Galilean satellites combined, compared to Jupiter. Also, the orbital distance of Titan from Saturn, in Saturn radii, is close to the median orbital distance of the Galilean satellites.[26] So Saturn and Jupiter have grossly equivalent systems, except that Saturn's is condensed into one big moon.

This has led to the hypothesis that something went *wrong* at Saturn—or right, if you're a fan of what you find! Several years ago Swiss astrophysicist Andreas Reufer and I proposed that Saturn once had a family of major satellites comparable to those of Jupiter. Instead of continuing to orbit in a coordinated manner, they once upon a time accreted into a single major moon, creating Titan and a handful of stragglers.[27] The mechanism we proposed is not too much different from how the Moon is proposed to have formed out of Theia, when it merged with the proto-Earth and left mantle remnants unaccreted, flung out as fragments.

The middle-sized moon Dione, in front of Saturn. The markings on Saturn are shadows cast by the rings, which can be seen edge on in the bottom of the image.
NASA/JPL/SSI

That's an origins theory, and therefore more speculative than a tidal evolution theory, which is on surer footing. From the estimated rates of tidal migration, dynamical modelers have been able to project the orbits of Saturn's satellites millions of years into the past, the way that Darwin projected the tidal migration of the Moon into the past, but with better data. These projections have shown[28] that Tethys and Dione and Rhea would have experienced powerful mutual couplings and probably collisions a hundred million years ago. If so, then nothing inside of Titan's orbit can be ancient. This is in agreement with our understanding of the rings, whose small mass should have been sandblasted away by micrometeorites on that same time scale. There must be an ongoing source to all this grace.

WHAT ABOUT TITAN itself? One puzzle with Titan is that its orbit is much more eccentric than any of Jupiter's Galilean moons. As we've noted, nothing comes for free: the dissipation (as heat) of orbital energy would have damped Titan into a more circular orbit long ago, shutting down tidal heating. To have an eccentric orbit today, Titan must have had an even more eccentric orbit a few billion years ago. Reufer and I found that a series of giant impact mergers could leave Titan with a 10 percent eccentricity, which would dissipate in time to its present value; in terms of tidal heating, it would be like storing wood for a long, cold winter.

Moreover, with a series of collision models, we showed that the diversity of MSMs could be explained, with tiny Enceladus coming from the lower mantles of the colliding bodies (hence its activity, decomposition of high-pressure phases) and watery Tethys coming from the icy exteriors. It became a geologically consistent story, if not yet a dynamically consistent one.

Titan's final accretion might have happened shortly after Saturn formed, or later as part of Saturn's early "jumping Jupiters" migration, or even later driven by satellite tidal migration. (A significant part of the story, I think, is that the Jupiter system avoided this fate. Jupiter is three times more massive, which would make its original satellite system much more stable.) Titan's formation by giant impact mergers would have left Saturn with one giant moon and a system of middle-sized moons that weren't quite stable—they would have a series of chaotic orbital problems to work out over the next few billion years, replenishing the rings now and then.

What can we conclude about all this? We have to go back to Titan! One, there might be life, and if so, it would have to be a second genesis, not a hitchhiker from early Earth. Two, it is the perfect analogue we need in understanding how climate affects a planet with a complex hydrosphere and atmosphere. Three, there is the rich geophysical puzzle about its origins that can tell us about the origins of Earth. And four, it is pragmatic: it would be relatively easy to land and operate space vehicles, since it has Earthlike atmospheric pressure and lunar-like surface gravity. But to voyage so far, and to potentially learn much, someone has to pay for it.

In the early 1970s, it seemed that the solar system would be an area of friendly competition among the superpowers, each one striving to outdo the other with technological advancements, so that we'd soon be exploring everywhere. Although the race to get to the Moon and Mars stalled for forty years, it seems to be waking up now that Chang'e-4 has landed on the lunar farside, where no spacecraft had landed before. With a modern human spaceflight program and a directed interest to explore the volatile-rich regions of the Moon's South Pole, China is paving the way toward an inhabited lunar base with in situ resource capability.

Meanwhile there is a private space race happening, with billionaires actively engaged in a private-sector competition: Virgin Galactic (Richard Branson), Blue Origin (Jeff Bezos), and SpaceX (Elon Musk), to name the most familiar, where the emphasis is on developing a market supporting space tourism, such as an Apollo 8–style flight around the farside of the Moon. I'd be on board with that. In all, twenty-six people have more wealth than half the humans on Earth and could easily send themselves on a comfortable landed mission to the Moon and back if that was their dream.

Fifty people are worth $10 billion or more, the cost of the JWST that is going to find the second-Earths. Thousands have private fortunes sufficient to fly an unmanned dirigible in the skies of Venus, land a boat on Titan, or return a sample of the plumes that are erupting on Enceladus. That none of them have done it demonstrates either a lack of interest or a lack of imagination or a lack of the awareness that it's feasible. Assuming it's the latter, and that you're the hundredth wealthiest person on Earth, let's suppose that you wake up one beautiful morning in one of your ten houses and have the epiphany to liquidate half your fortune. You've decided to finance an advanced mission of solar system discovery that would be one for the history books, right up there with Voyager and Viking and Cassini. You can do it. Where would you go?

If you have a seafaring passion, choose Titan. Beautiful and geologically complex beyond belief, it is the only attainable planet besides the Earth that has a massive atmosphere and open seas. Apart from being muggy and hazy in its cryo-organic way, Titan offers a comparatively benign environment for a creative, ambitious mission. Using data from Cassini and its Huygens lander, your team could minimize operational risk while preparing for the planet's unknown intricate marvels and surprises. NASA just selected the Dragonfly mission, a refrigerator-sized quadcopter that will land in the dark dunes of Shangri-La in 2034 and then buzz around for a hundred miles or more.[29] It will descend into an 80-kilometer diameter crater and map the geology and sample the molecular and isotopic composition. Coming in with several times the budget, you could get there faster; if you overlapped, your data streams would have to be coordinated through the Deep Space Network.

Another advantage of adventuring on Titan, compared with Mars, the other spectacular Earth analogue, is that hardened microbial stowaways on board flight hardware, which are impossible to sterilize completely and can survive voyages in space, would be unable to proliferate on the cryogenic, methane-saturated surface of Titan. Unlike a mission to Mars, where some terrestrial organisms could probably thrive in certain niches, a Titan mission is safe according to the regulations of *Planetary Protection*—the international agreement that protects possible alien life from being wiped out by Earth life before we have a chance to discover it. This prudent long-range view can make missions prohibitively complicated, expensive, or even impossible—a fact that prevents us from visiting the most geobiologically interesting areas on Mars with current technology.[30] You don't have to worry about any of that on Titan.

But let's not get ahead of ourselves. Early on, with a chief engineer whose opinion you trust, you will decide on a launch vehicle that is sufficient to get you to Saturn as fast as possible with a two-ton spacecraft. That decision will define the trajectory and mass parameters for your mission and get you there in around five years. You will also decide on a spacecraft propulsion system, whether that be chemical (rocket fuel) or ion drive (solar or nuclear[31]), teaming with a vendor or with a space agency who has flown this sort of thing. You would contract out mission design, launch, and flight operations to a proven deep-space mission center; the choice is limited, given their need to be familiar with hypersonic atmospheric entry. Prepare to write them big checks, and what you get in return is the assurance that your mission will be designed, built, flown, and managed according to the best practices. Additionally, because Titan gets only 1 percent the sunlight that the Earth does and is shrouded beneath an atmospheric haze, the power for mission operations on the surface will have to be nuclear—e.g., a radioisotope thermoelectric generator (RTG) that converts heat from plutonium into electricity—so it's prudent to have a spaceflight center procure the power pack as well.

Now that you've got a ticket to ride and a launch date to meet, and a spacecraft to get you to Titan, you can focus on the exploration itself—what you will do and how. That begins by building a core science team, starting with people you'll enjoy spending the next ten years of your life with, whom you enjoy solving problems with and whom you trust to make sound, timely decisions—people who will treat you as a peer and push back when they think you are wrong. Recent missions have proven that diverse teams, in age, race, and gender, are the most scientifically successful.

The first task will be to come up with the concept of operations (conops). You want mobility, but will it be wings, legs, balloons, pontoons? A sailboat? Part of that decision will depend on the science payload: the instruments, cameras, chemical sensors, manipulating arms, and radars. The chief engineer will oversee the payload as it fits within the mission architecture, especially as to how it lands on Titan, what it does on the surface, and how it gets that data back to Earth. Her job, especially during the final design, is to tell you the bad news if the mission is projected to go over budget or falls behind schedule, and then come up with a detailed plan and possible descopes to get on track.

You've long enjoyed sailing and decide that it would be great to pilot a boat around Titan and make radar and acoustic soundings of its lakes and rivers as well as laser measurements of cove and cave topography. You want to measure detailed seafloor bathymetry and look for macroscopic organisms like cryo-squids and fish, and even reefs, dragging a seismic and radar imaging array behind the boat. To get some ideas, you've been boating with friends in Scotland, where the highland lakes are comparable in depth and topography to some of the lakes of Titan, testing out final prototypes of the vessel and its instruments. The decision is made to focus on highly detailed yet basic measurements—visible pictures, laser scans, images of structure, microscopic imaging and chemical analysis of the fluid—a two-year mission.

A parachute drop into a lake could deliver substantial mass to the Titan surface, in the form of a nuclear-powered boat. The basic

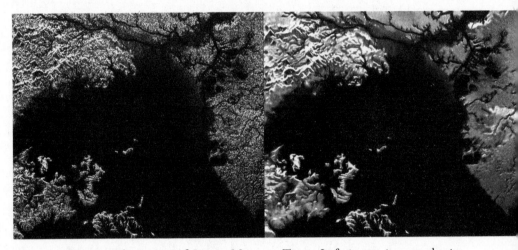

Cassini radar image of Ligeia Mare on Titan. Left image is a synthetic aperture radar (SAR) image; the radar penetrates tens of meters into the methane-ethane seas to reveal near-shore bathymetry. Right image is processed for electronic noise (despeckling). Ligeia Mare has a span of about 500 kilometers; shown is the southern half, where Vid Flumina (to the top right) connects seasonally with the largest lake, Kraken Mare to the southeast.
NASA/JPL-Caltech/ASI

idea was first proposed by American scientist Ellen Stofan in 2009, whose team designed an instrumented buoy that would drift in the slow currents and meter-per-second winds. With three or four times the budget, your boat will be an inflated pontoon craft with a nuclear power source in the keel that could last for more than a decade, providing ample electrical power for science instruments and communications, and for heaters to stave off the cold.

After looking over large-format radar and visible-infrared images from Cassini and Huygens and listening to days of presentations by your science team, you decide on Ligeia Mare, which is where Stofan's team was going to land.[32] It is comparable in size to the Great Lakes but much shallower and filled with hydrocarbons, fed by subterranean aquifers and a seasonally flooding river. Vid Flumina extends hundreds of kilometers into the eroded uplands, through canyons carved into ice, and its river mouth will be your goal. You want to be prepared for currents and possible winds, and thus your vessel needs to be stable, maneuverable, and unsinkable, with reliable propulsion.

The spacecraft will be about the size of a large van, with a heat shield to survive the dive into Titan's atmosphere at 9 kilometers per second. That sounds scary, but it works just fine, since your drop zone is the size of Lake Superior and the atmosphere is stable and thick. Inside the heat shield will be this incredibly specialized boat that has been designed based on Mars and Moon rover designs. It uses systems that are proven to work, the main novelty being the pontoons and propellers. As for operations, due to the great distance the vessel will have to operate autonomously, with commands uploaded only every couple of days. It will execute a sequence and then report on it; you will get a travelogue from the boat an hour after the fact.

Once the mission has launched and is in space, you can enjoy the next five years relaxing and tending to your remaining fortune and doing whatever it is that billionaires do, hanging out on some James Bond island. There will be a couple of mission-critical events during the years-long cruise to Saturn; these will be fun since there are seldom any problems. (Invite the neighbors.) There may be an asteroid flyby, and maybe one or two planetary close encounters for gravity assists where you'll obtain some images of worlds en route. There would also be a couple of midflight trajectory correction maneuvers (TCMs), in which a small thrust is fired to get it perfectly lined up for Titan, where the atmospheric entry has to be precise. The positive publicity of these successful and exciting mission events might help you recoup the billions your friends said you had squandered on some weird thing in space, the way the curmudgeons complained that the Apollo space program was costing billions, when it stimulated the economy a hundredfold.

The laws of physics, combined with exacting tolerances, make modern spaceflight a miraculous success. It all seems so audacious, landing a boat on a satellite of Saturn, so you just have to pinch yourself as the nerve-racking moment approaches, as your lander readies to slam into Titan's atmosphere, using the heat shield to stop the spacecraft from its interplanetary trajectory. Once it's lost most of its incoming velocity, the drogue chute opens, which in turn pulls open the main chutes. All your hope is in the preprogrammed electronics sequences and algorithms, the mechanical deployments and control systems. Plenty can go wrong, it seems, but little can *really* go wrong if the mission is designed right. If your team is good they will have thought of everything, and will have beaten down the risks with redundant engineering, and by studying every scrap of previous mission data for planning.

As these spacecraft events come down via telemetry and are decoded, people in the control room exhibit great anxiety and excitement. There is loud applause when confirmation arrives that the pontoons have inflated, that landing is imminent. But here's why it's all so weird: the landing actually happened almost an hour ago, a billion miles away at the speed of light. Your spacecraft has either deployed or is scattered on the surface in broken pieces. Schrödinger's cat is alive or dead,[33] and nothing can be done. If you wanted to fix anything anomalous one hour before landing, you couldn't, because you exist in *its* past. By the time it receives your message, it is busy with its initial checkout, or else it is reduced to a few scraps of debris snagged on a floating parachute, drifting toward the islands by the shore.

Let's say you've landed close to your targeted location, 16 kilometers from shore, and your boat has deployed without any significant anomaly. Its self-rigidizing structure now supports a platform of science instruments, a large radio antenna for communications with Earth,[34] and beneath all of that, the radioactive thermal electrical generator. Sensors and imagers are distributed around the platform for virtual reality data products, as well as radars and sonars peering around and below. The small microscopy and chemistry lab will begin sampling and analyzing the hydrocarbon lake. Computers onboard will reduce the raw data into a selection of radargrams and sonograms, and mpegs and so on, instead of trying to transmit all the raw data to Earth.

Seven days later, the hoopla of the landing is behind you, and your team has completed its first week of deployment, motoring steadily through the calm waters toward the island-studded coast. In the Science Operations Center (SOC), there are maps and globes and wall-sized computer screens for projecting images. So far the sea is calm as glass and the wind is blowing at about one knot. It sometimes blows hard enough on Titan to pile up sand dunes, but for now the planet's weather is cold and sluggish, and as always, hazy. After four days of constant motoring, the pontoon stops and waits for further instructions, as it has just transmitted the first images of the islands and small crenulated bays that line the icy shore. You see a bedrock of ice that has been eroded, exposing caves. Embedded in the sea walls are ancient layers revealing the planet's deep geologic record.

In a few more days you reach the river mouth, the target of your primary investigation. Your boat starts making monotonous back-and-forth radar and sonar surveys that paint a comprehensive and very detailed image of the channels and deltas and other submarine landforms. Around the world the mission has generated endless fascination, but in the SOC you "live" on Titan and there's too much work to notice. Plus, you've just been delivered the first partial installment of VR data. You head into the soundproof studio behind the SOC, walk to the middle, and push a button that turns on a cylindrical wall of color monitors, illuminating you in a glow. Then you turn on the volume and hit play. The screens come to life with vivid color images stitched together into an enveloping expanse. You listen to quiet sounds recorded earlier that week by your spacecraft's microphones—raindrops, wind, the lapping of "water." You have set up little fans and sensors around the studio that blow air to match the pressure sensor and anemometer data; now you even *feel* like you are there. New ten-minute loops will be transmitted every week.

As the floodgates of data open, your team pores over it all, confirming or denying old theories, coming up with new ones. After a period of about six months, you start to stream the raw data into the public domain, and team members deliver validated information and processed data products to the permanent archives at NASA and ESA. Having completed your primary mission objectives, your team pilots around the shores looking for signs of unusual geological activity and searching for evidence, against all odds, of past or present life. If you do find life, then this story takes a completely different arc and does not end. Otherwise, you now begin the long-planned higher-risk adventures. At the mouth of Vid Flumina, where you have mapped out the gravelly shallows, your boat makes a beach landing and deploys a small and simple rover that drives up into the gravels and rocks—hard ices, wet with hydrocarbons, blackened with organics—and looks back to snap a picture for us all to see: a boat in a bay, a billion miles away. Then it would tool dutifully ahead until it got stuck or boxed in or ran out of battery power. You'd sail on, piloting your boat up the river like Humphrey Bogart in *The African Queen,* no turning back. Would you discover life? Of course you would: inside yourself.

CHAPTER 5

PEBBLES AND GIANT IMPACTS

Should a Comet in its Course strike the Earth, it might
instantly beat it to Pieces, or carry it off out of the
Planetary System. . . . Disputes between the Powers of
Europe would be settled in a Moment.

—BENJAMIN FRANKLIN, *POOR RICHARD IMPROVED,* 1757

I
T IS OFTEN EXPLAINED THAT the asteroids orbiting the Sun
and the meteorites that derive from them are leftovers of the
planet-forming stuff. That is misleading. Think of a planet as
a finished house and a solar system as a subdivision that goes up.
Then the original stuff is the pallets of wood and gypsum board
and roofing sheets, the sacks of rebar and bins of nails and screws
and bags of concrete delivered to the job site. Asteroids are more
like the construction scraps spread out on an empty lot after the
subdivision is completed. Sure, you'll find some original stuff, but

mostly you'll find sawed and painted lumber and ripped-up bags, roofing scraps and chunks of drywall, stacks of broken pallets and heaps of wire. Studying asteroids and meteorites is like sifting through the rubbish bins in the dark to find out how your house was built.

Asteroids represent a mysterious region of the solar system. If the original protoplanetary disk extended uniformly beyond the terrestrial planet region, without a gap, then you would expect a several-Earth-mass planet between Earth and Jupiter. Instead there is Mars, at 1.5 AU, and the Main Belt of asteroids, then Jupiter. This jarring transition between the tapering terrestrial planet region and the outer solar system is unexplained, and is the primary motivation for the Japanese space agency's forthcoming Mars Moons Explorer (MMX) mission, to land on Phobos and return a sample.

Looking sunward from Jupiter is like standing on a precipice or swimming out beyond a steep ocean reef—there is little matter to speak of, a few asteroids from 2.4 to 3.5 AU adding up to a few percent of a lunar mass, then Mars. There is also a distinct compositional gap in the rocky components of the solar system; geochemist Paul Warren of UCLA has shown that carbonaceous chondrites came into the picture late, from the outer solar system leading to ideas that Jupiter formed around a rocky core and established a grand compositional divide.

In any case, our solar system appears to be anomalous, and these structural and compositional gaps could be an indication of what happened. Most systems discovered around other stars are much more tightly packed than ours, with Jupiter-sized planets well inside of 1 AU. There is a strong selection bias because it is far easier to detect planets that are massive and closer to the star. Still, considering the thousands of planets that have been discovered and accounting for these biases, it appears that most solar systems have inner giant planets hundreds of times more massive than the

Earth, "hot Jupiters," a collection of weird monsters with orbital periods of days. Only a small fraction[1] of solar systems have outer giants comparable to Jupiter or Saturn.

Most planets discovered so far are "super Earths" and "mini Neptunes," around 3 to 10 Earth masses. It is unusual for there to be no planets of this size in our solar system; we have to infer what they are like from theory and very limited data. Although some appear to be in their star's habitable zone, most are well inside the equivalent orbital distance of Mercury (0.4 AU). As for Earth-mass planets orbiting around 1 AU from a sunlike star— Goldilocks planets—here there is a wide blank region on the plot. The only *known* planets of this kind are Earth and Venus.

Something has made our system the way it is, an event or epoch that may be fossilized in this gap between the Earth and Jupiter. There's Mars, which is dozens of times less massive than you'd expect, and the asteroids, which are depleted by at least a factor of 1,000 relative to what once was. We get a sampling of Main Belt asteroids for free when they drift into near-Earth space. Resonance forcing by Jupiter or Saturn torques their orbits, and the thermally driven Yarkovsky effect causes small asteroid orbits to migrate,[2] shuffling them around until they scatter into near-Earth space. A related thermal effect called YORP can cause asteroids smaller than a few kilometers diameter to spin up like pinwheels, faster and faster; if unchecked, such a pinwheel can break in two and fling off a moon—not far off from what Darwin envisioned but at 1:10,000 scale.[3] And of course, asteroids get pummeled by smaller asteroids, making still smaller asteroids, and ultimately the scraps from the boneyard known as meteorites.

THAT'S THE MODERN picture for interpreting the cosmochemistry of meteorites: they are samples from the Main Belt, or what's left of it, for the most part. But until the 1800s meteorites were thought to be atmospheric phenomena[4] (Greek *meteoron* = out of the air). They have been venerated as sacred objects throughout history and pre-history, and iron meteorites have been used to make knives and daggers throughout the world. One early scientific idea for the origin of meteorites was that they were erupted out of volcanoes and flew hundreds or even thousands of miles until crashing back down. In 1864, a primitive black stone exploded into fragments high above the town of Orgueil in the Pyrenees, and fragments weighing up to 14 kilograms were recovered before the rest of the soft object disappeared into the soil

The Earth rams through the tail of Comet Tempel-Tuttle in this woodcut of the famed Leonid meteor shower of 1833 that startled everyone but caused no damage and produced no meteorites. The effect is like driving into a snowstorm.
Woodcut by Adolph Völlmy (1889)

and vegetation, becoming part of the earth. Orgueil was the first meteorite subject to widespread scientific attention, in part because analytical science was ready for it, but in part because the fresh fragments were so strange and smelled like peat, like organic matter. Where did it come from?

The British mineralogist Henry Sorby argued in 1877 that chondrules were perhaps "droplets of fiery rain" ejected from the Sun in prominences,[5] which fit in with Kelvin's idea,[6] popular at the time, that the Sun is "an incandescent liquid now losing heat." Or else meteorites maybe come from the Moon, launched from its craters, which were thought to be volcanic. For the next century the debate stirred quietly among those who thought meteorites were related to terrestrial phenomena and those who thought they were rocks from space,[7] and if from space, whether from the Sun or the Moon, and if the latter, whether by volcanoes or impacts—or from somewhere else, however weird that might seem.

Shortly after Orgueil fell, the Italian scientist Giovanni Schiaparelli—who would later be more famous for his cartography of Mars, introducing the concept of *canali* for Lowell to fixate upon—established the first concrete dynamical relationship between comets and meteors. He showed that the Perseid meteor shower, a brilliant show that happens every August, coincides with the Earth crossing the eccentric orbit of comet 109P/Swift-Tuttle.[8] Comets, he claimed, fall apart and become the sources of shooting stars (and he was right; we run through their fragments like a car splatting bugs on the windshield). Does this mean that comets are the parent bodies of meteorites? If so, why don't meteorites rain down from the sky during meteor showers? It still did not make sense.

If you open up a sealed canister containing fragments of Orgueil, you can smell what for me is a faint trace of crumbled egg yolk and turpentine. Crush it in a mortar and you can experience its grainy, crumbly texture, like dried-up clay. Primitive meteorites like Orgueil are *carbonaceous chondrites*, a relatively common class of meteorite; of these, the most volatile rich (that is, with compounds like water and hydrocarbons) are extremely precious and must be collected shortly after they fall. This does not mean that they are precious in space; many NEOs could be like Orgueil, and the OSIRIS-REx mission target, asteroid Bennu, seems to be even more primitive than that. They're rare on Earth because they are so fragile that they explode high in the atmosphere; sample return missions are required if we are to study the most primitive materials.

Montage of three near-Earth asteroids, each of them the target of a sample return investigation. The small one, 350 meters long, is Itokawa, subject of JAXA's Hayabusa-1 mission that completed (barely!) in 2010. Next in size is Bennu, 500 meters diameter, blackest of the three, with a really mysterious and complex surface, probably more primitive than any meteorite that's made it to Earth. (We're currently looking for the right place to grab samples on Bennu.) The "big" one is 900-meter Ryugu, where JAXA has brought sub-spacecraft and landers along for the ride, and will also retrieve samples. *JAXA/ISAS; NASA/GSFC/U. Arizona*

The largest carbonaceous chondrite meteorite ever found exploded over the state of Chihuahua in northern Mexico in February 1969, only months before the first humans landed on the Moon. Tens of tons of its fragments were strewn around a 400-square-kilometer expanse near the village of Allende, and enterprising cosmochemists from laboratories in the United States drove down and recovered tons of it. Many lunar scientists at the time felt that carbonaceous chondrites might come from the Moon; this was in part because the bulk density of the Moon, known since the 1870s to be about 3.3 grams per cubic centimeter, is comparable to the bulk density of these stones. Until the *Eagle* landed and proved otherwise, prominent scientists maintained that the Moon's dark maria were covered in a fine carbonaceous powder, and there was a valid concern that the astronauts would sink into thick dust. By means of state-of-the-art equipment ready to analyze the first Apollo samples, Allende became the best-studied rock on Earth.

As luck would have it, another large carbonaceous chondrite fell in Murchison, Australia, a few months after Apollo 11 had returned with its boxes of igneous rocks. Murchison is much richer in complex organic molecules than Allende, a cosmochemist's delight, and is far more interesting than Moon rocks in terms of what would become the emerging science of *astrobiology*. About 100 kilograms of Murchison were recovered, making it a reference standard for primitive composition and allowing for comprehensive chemical analysis of its constituents. Most eye-opening from the study of Murchison was the discovery of dozens of amino acids.

Amino acids are the simplest building blocks of proteins and thus connected to the emergence of organic life. Amino acids have also been discovered in the dusty tails of comets, as determined by their spectral absorptions. Not only is Murchison chock-full of amino acids, but more than two-thirds of those molecules are "left-handed"—that is, its organic molecules have a *chirality* like the

So far, there are no known meteorites from Mars that have a sedimentary structure, but we know they are there. This view of Vera Rubin Ridge shows us the kinds of rocks we'd like to sample. Named for the American astrophysicist who discovered the evidence of dark matter, the ridge was imaged by the ChemCam on NASA's Curiosity rover. Stacks of sedimentary rocks have been fractured and jointed, and precipitated minerals are exposed, left behind by the circulation of underground fluids long ago. The front of the ridge is about 5 meters wide in this view.
NASA/JPL-Caltech/CNES/CNRS/LANL/IRAP/IAS/LPGN

fingers of your left hand. Point your left thumb toward you, and your fingers curl clockwise; that is left-handed chirality, which is opposite the curl of your right. Other examples of left- and right-handed chirality are the screw threads on your left and right bicycle pedals, which have exactly the same function but are not interchangeable.

Every living organism on Earth has left-handed chirality in its DNA. If life had evolved with a right-handed chirality, we would be no different functionally, but we cannot utilize chiral-opposite molecules; fats and proteins of right-handed chirality would be of no use to us.[9] Many organic compounds in addition to amino acids are found in the most primitive carbonaceous meteorite: aliphatic and aromatic hydrocarbons, fullerenes, carboxylic and hydroxycarboxylic acids, purines and pyrimidines, alcohols, and sulfonic and phosphonic acids. It sounds like the back of a bottle of health supplements. (While I don't recommend eating meteorites, I know of at least one instance of an adventurous cook using them in small quantities.)

Is this left-handed chirality coincidence or is it causal? Did we inherit the blueprint to our DNA from asteroids and comets? If so, they are indeed the "sous chefs" of life, working in the irradiating kitchen of space, preparing the mixtures and broths from which life would originate. And if that's the case, then these most basic preparations for organic life are everywhere in the Universe!

WHEN AN ASTEROID or comet passes near a planet, this has a major effect on its orbit. (Likewise, since every action has an equal and opposite reaction, the encounter will nudge the planet ever so slightly—the cause of giant planet migration.) Ultimately these encounters will randomize the orbits of the small body population, increasing their eccentricity and inclination, scattering them until one of them hits the Earth—and everybody dies. That is the danger of planetary chaos in a nutshell, the earliest description of which is by Isaac Newton in the second edition of *Opticks* (1706):

> For while Comets move in very excentrick Orbs [. . .] blind Fate could never make all the Planets move one and the same way in Orbs concentrick, some inconsiderable Irregularities excepted, which may have arisen from the mutual Actions of Comets and Planets upon one another, and which will be apt to increase, till this System wants a Reformation. Such a

wonderful Uniformity in the Planetary System must be allowed the Effect of Choice.

Why hasn't one hit us yet? "For it became Him who created
them to set them in order." In other words, God has been running
interference, looking ahead to make sure things aren't going to
crash into us. Gottfried Leibniz scoffed, in a 1715 letter to Samuel
Clarke (a follower of Newton whom he was trying to convince),
that in Newton's philosophy "the machine of God's making, is so
imperfect [. . .] that he is obliged to clean it now and then by an
extraordinary concourse, and even to mend it." God is the divine
clockmaker, but then he is also the clock repairman—and a bad
one at that.

Chaos makes the orbits of small bodies unpredictable on time
scales of decades to centuries. Once they get entangled in Earth's
gravity, their irregularities are "apt to increase, till this System
wants a Reformation." Newton's thinking is remarkable for having
appreciated not only the inevitability of chaos, but also the "Effect
of Choice" in moving celestial bodies around to better locations.
Today the effect of choice is in our own hands, rather than in the
hand of God—although that is not excluded—and someday, perhaps hundreds of years from now, a space mission will be sent by
the agencies of the world to deflect a threatening near-Earth asteroid that is on a collision course. But before we can do anything
about them, we need to understand their physics.

NEOs represent important and accessible science mission targets, voyagers from many different parts of the solar system brought
within reach. But they also represent a hazard that has a real probability of global consequences in the next few thousand years.
Once we are able to modify the orbits of hazardous comets and
asteroids, maybe we'll be like gods and use our powers to create
celestial fuel stops, water reservoirs, iron and platinum mines, and

platforms for comfortable space colonization. We could go beyond being clock repairmen to improving the workings of the clock, putting dangerous asteroids into better orbits, sending some of them that have lots of resources into lunar orbit (moons of the Moon), and adapting one or two to cycle continually between Earth and Mars, using their regolith as radiation shielding[10] for voyages lasting a year or more.

TO THE SURPRISE of many, when we first saw asteroids up close in the 1990s, none of them looked like mere hunks of rock. Had Robert Hooke ever seen an asteroid up close, he would have observed, as he did for the Moon, that they exhibit a "principle of gravitation, such as in the Earth" that holds them together and makes ponds of dust and gravel and layers of conglomerate materials. They don't have much gravity—just enough to account for geologic structures that look like canyons and mesas and dunes, and impact craters with massive rims, and huge boulders with fragments scattered all around them.

Asteroid landscapes are fantastic, their rocks sitting cartoonlike on ridiculously steep perches. Most of them spin much faster than the Earth, some to the point of catastrophic breakup. Many of them have top-like shapes with equatorial ridges, where the "low" (in terms of where a ball would roll) is the high ground in terms of elevation from the center. For some small asteroids, gravity is close to zero at the equator, when you factor in the centrifugal force. (There's still gravity, but you're basically in orbit.) It's all kind of crazy, and that's what makes landing on an asteroid—seemingly a simple, gentle task, sort of like docking with a space station—potentially complicated and even dizzying.

The first time humans created artificial gravity in space was in 1966, when the Gemini 11 astronauts attached a hundred-

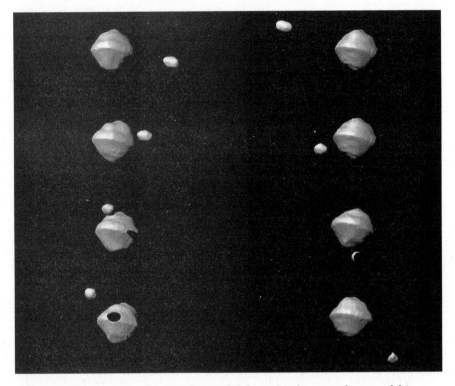

Shape model for asteroid 1999 KW4, the first clearly imaged asteroid binary system, constructed by American astronomer Steven Ostro and his colleagues using high-resolution radar data from Arecibo Observatory in Puerto Rico. Shown to scale are the primary body, Alpha, and the satellite, Beta. Alpha rotates with a period of 2.8 hours, while Beta is tidally locked. The ridge and even the moon may have been consequences of previous spin-up to even faster rotation.
JPL Digital Image Animation Laboratory

foot tether between their capsule and a vehicle that was kept in low-Earth orbit for docking practice. They were getting ready for the Apollo missions. By thrusting against the tether, they carefully initiated a co-rotation of about one-sixth of a revolution per minute, enough to create a centrifugal force equal to gravity on a 10-kilometer asteroid. The slow bolero was too subtle to be felt by either astronaut, but they noticed that a camera slid across one of the instrument decks. (Actually the camera tried to maintain its momentum, while the spacecraft was tugged by the tether in a different direction.)

Now imagine they'd opened a bag of coffee beans inside the capsule during this slow spin. At first the beans would appear weightless, adrift. But given enough time they would pond to the "bottom," just as they would if the capsule was at rest on a small asteroid.[11] Movements are slower than the minute hand of a clock

Not the Moon. This is Sedan Crater, and other craters, of the Nevada Test Site. When a nuclear bomb (in this case, 104 kilotons) is buried the right distance beneath the surface (in this case, 285 meters), it can have a very similar formation mechanism and a close geological resemblance to a celestial impact event of equivalent kinetic energy.
U.S. Department of Energy

and too subtle to feel, so the beans might not settle until the astronauts had slept for a couple of hours. This is why it's good advice not to do anything too fast when you're on an asteroid; if you try to dig a hole, you might end up creating a landslide or even a dust atmosphere that would not settle for days. An asteroid is like a clear, quiet lake bed, where it's easy to stir up the muck.

Asteroids are so completely yin in terms of their dark and subtle geology, yet they can cause the most violent geological transformations possible when they impact planets. The most recent large airburst was above Chelyabinsk, Russia, in 2015, when a 20-meter stony meteorite detonated across the sky in an explosion equivalent to thirty Hiroshima bombs. Major fragments were recovered. The largest scientifically documented meteor was a century before, in 1908, when a 30- to 50-meter diameter asteroid exploded 5 to 10 kilometers above Tunguska, Siberia, an area that is famously remote. Ground zero wasn't visited for almost twenty years, with the tsar's wars and all; the first expedition in 1927 found over 2,000 square kilometers of flattened forest with small trees growing back, and collected samples and accounts from a few eyewitnesses. But no meteorite fragments or undisputed chemical traces were ever found,[12] leading to alternative explanations such as (of course) an exploding alien spacecraft or a mini black hole.

Smaller events are much more frequent, and it might shock you to learn that Earth is struck by a meter-sized meteoroid every few weeks, producing a 1-kiloton explosion. (The Hiroshima explosion was 15 kilotons.) Real-time monitoring of meteoroid airbursts—small NEOs—became accessible to astronomers in the mid-1990s, when the Department of Defense grew concerned that something was wrong with their sensitive new detectors flying on spy satellites. They were observing what appeared to be bombs going off in random locations. These are the flashes of meteoroids, and scientists have gradually gained further access to this treasure trove of data: sizes of NEOs, airburst heights, and compositions, detected from space and from the ground.

A small asteroid impact is not all that different from a weapons test, and this leads to another connection between the cold warriors and the astronomers, understanding the physics of these events. Some of the best computer codes for simulating asteroid impacts, and the fastest computers to run them on, are maintained at nuclear defense research labs, which relish the opportunity to validate their codes. Also, these labs have taken an increasing interest in the physics of disrupting or deflecting an asteroid or comet if one is ever found to be on a collision course with Earth.[13] One approach is to detonate a nuke about one asteroid-diameter away, so that the X-ray flash heats and vaporizes the surface rocks on one side, generating an impulse that nudges the asteroid in the opposite direction.

If this connection between astronomy and the nuclear arsenal seems strange, recall that it was solving the trajectory of the cannonball that motivated Galileo's laws of motion, not the orbit of the Moon. The first telescopes were coveted far more for their use in sea battles than for gazing at the small, bright stars around Jupiter. Rockets that take research payloads and astronauts to space were invented in World War II and perfected in the cold war for launching ICBMs that could destroy whole cities; voyages to the lunar surface were incidental. The military spends just as much on space telescopes as NASA does, but theirs are pointed down. Modern telescopes use adaptive optics, a technology that grew out of research to acquire detailed images of enemy satellites passing two hundred miles overhead. Radio astronomy flourished with the commissioning of large military radars during and after World War II. The first images of the lunar farside were obtained using the best Russian spycam technology and purloined American film. Strange bedfellows.

THE OUTER SOLAR system, beyond Jupiter, has two populations of small bodies, those that were there to begin with—original icy objects that never participated in planet formation—and those that got ejected when the giant planets formed and jostled for position. The formation of the giant planets cast out an estimated one trillion comets that would become the Oort Cloud. A hundred billion more probably "went the extra mile," escaping the Sun, and are now on their way through interstellar space. And billions of years from now, when the Sun loses half of its mass, it will lose hold of its outer Oort Cloud, and trillions more objects will disperse throughout the galaxy.

However they got there, the small bodies of the outer solar system have been in cold storage for billions of years, which makes them exciting targets for spacecraft missions to determine the starting conditions of planet formation. The most distant bodies have surface temperatures only tens of degrees above absolute zero—that is, over 200°C below zero.[14] Whether they are cold deep inside depends on whether they are big enough and rich enough in rock to sustain radioactive heating, and whether they end up as large satellites in multiple systems that can have significant tidal heating. Pluto and its large moon Charon, for example, would have induced tidal heating inside of each other for the first hundred million or so years, until they both became tidally locked; now this heat source is gone.[15] There may lurk giants farther out there, and Planet X could be an Earth-mass binary planet for all we know. But most of the cometary objects that we know of are far too small to have experienced any substantial heating of any kind, radioactive or tidal—hence the importance of getting samples of their truly primeval materials.

Occasionally one of these outer solar system bodies works its way back in. A distant comet in the Oort Cloud may experience a galactic tide when its orbit is at its farthest from the Sun, thousands of AU away, causing its orbit to bend a little so that its next orbit is slightly offset, until at some point it encounters the gravity field of Neptune and finds itself in trouble, like a moth getting caught in a web. Some of them get tangled up with Jupiter, and most that do don't survive very long. Brave ones called *Centaurs* are scattered KBOs that are in the process of trying to get past Jupiter; they exhibit all kinds of unusual and unpredictable activity as their ices feel sunshine for the first time ever.

Comets are made of dust and ice, including *super volatiles* such as CO, CO_2, N_2, CH_4, and O_2, molecules and compounds that vaporize at extremely low temperatures. With minimal heating they start to come apart energetically, producing water vapor and silicate dust. They undergo sublimation-degradation as their cementing materials vaporize and go away. Amorphous solids deep inside the comet—ices, except they condensed directly from the nebula 4.6 billion years ago and never got *warm* enough to crystallize—transform into crystalline ice in a reaction that might be *exothermic,* giving off heat and causing the wholesale decomposition of the nucleus.

Some comets die in the attempt to make it through Jupiter's walled gate. Some crash into Jupiter; others are scattered by its powerful gravitational influence. Still others are disrupted by the giant planet's intense tidal field, forming a dozen smaller comets during a single close flyby (something that happened to comet Shoemaker-Levy 9 in 1992). Once past Jupiter, they are called JFCs (Jupiter Family Comets) and are thrown onto chaotic orbits that can collide with any of the terrestrial planets, or fall farther inward, until the Sun breaks them up by tides and radiative heating until nothing but dust is left.

Such is the birth, life, and death of comets—a frenetic first hundred million years, then four and a half billion years in cold storage, and then, for a lucky few, an exciting phase lasting tens of thousands of years when they plunge sunward for a grand finale. A comet's orbit may grow eccentric and end up well inside of 1 AU at perihelion, although spending most of its time far beyond Neptune. Every perihelion, every several thousand years at first, would expose its increasingly coal-black surface to the Sun, which would heat it to hundreds of degrees, like pavement in July. This heat takes time to propagate inward, and the surface temperature oscillates between hot and frozen as the comet rotates (day and night). The near-surface materials become highly processed under this rotisserie, becoming an ice-cemented carbonaceous lag on top of a primeval interior.

My mother once made a dessert called Baked Alaska. She put a block of frozen ice cream into the oven on high for several minutes, until the outer layer caramelized while the interior stayed frozen. (You had to bake it on a wooden board, because a glass or metal pan would conduct heat and melt it.) The holy grail of planetary geochemistry is to get inside the caramelized skin of that dessert, to get at the frozen ice cream, which would be untouched material from a body that, only a few thousand years ago, had been in storage beyond Ultima Thule. These pristine solar system solids are now buried under meters of evolved lag—stuff that doesn't sublimate away—including the organic molecules that make a comet black. Still, even an "old" JFC that has suffered dozens of perihelia is likely to have fresh, easily samplable surfaces, given how geologically active these bodies are. Comet 67P/C-G has newly formed cliffs and crevasses, exposing fresh interior materials waiting to picked up. The CAESAR mission was therefore proposed to the NASA New Frontiers program to go back to 67P and collect its precious treasure. But after fierce competition the Dragonfly mission was selected instead, to be launched in 2026; if all goes well it will begin hovering around Titan's landscape in 2034.

Great comets are rare, but they don't have to be especially large—they just have to pass near the Earth and throw off dust trails and comas. Small comets, being closest to their final days,

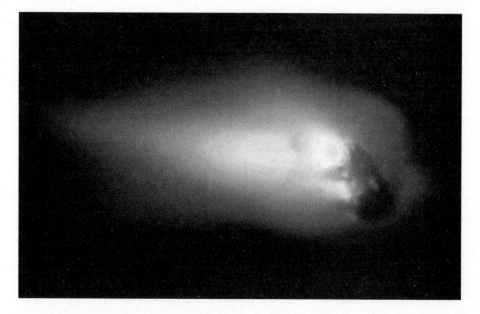

In 1986, the Giotto spacecraft passed within 600 kilometers of 1P/Halley as the comet zoomed away from the Sun at a record flyby relative velocity of 68 kilometers per second. The coma is blown downrange by the solar wind. The spacecraft was hit by a small grain and sent into a tumble, and its camera was destroyed, but not before sending back this first geologic image of a comet nucleus, 16 × 8 kilometers diameter.
Halley Multicolor Camera Team, Giotto Project, ESA

can put on the biggest show. And there is the element of luck and position—a great comet needs to be high in a moonless sky to be truly spectacular. I have had the luck to see two great comets. For one of them, Hyakutake, I was camped out with my partner in the darkest desert, twenty miles from the smallest town. We lay back in our sleeping bags and watched the night come alive. The comet came out like an opera singer beginning her aria, and by midnight she was overhead. I relaxed my vision and opened my senses to the nucleus interacting with the Sun, its surface hot for the first time in its life, its dust and ice erupting and getting blown away by the solar wind, and its glowing, strongly ionized tail.

Only a tenth of an AU from Earth at closest approach, and with an exceptionally long tail, Hyakutake moved appreciably relative to the stars from hour to hour through the binoculars. To me it seemed alive.[16] The scene was so sweetly moving, full of greens and blues and creamy whites, that I didn't know if I was sleeping when I woke, hours later, to the cold night. The hills to the east were etched with cobalt blue, hints of dawn. I saw a few shooting stars and reflected that I'd never again see anything so beautiful. I dozed and woke, and now the morning was washing it away, to exist in memory. I don't know the next time I will see a great comet place its hush upon the land, so I live where it is dark.

The most famous comet is 1P/Halley (P for periodic, 1 for 1st), the object that proved Newton's laws. In 1705 Edmond Halley published calculations showing how the great comets of 1531, 1607, and 1682 were (he proposed) one and the same comet, with aphelion (farthest distance from the Sun) just beyond Neptune and

Artist's concept of comet 1I/'Oumuamua. It was not imaged, but is inferred to be a dark red, highly elongated object, estimated at 230 meters long and only 35 meters wide. Slightly tumbling, it makes three revolutions per day. *ESO/M. Kornmesser*

coming back approximately every 75 years. Applying Newton's law of gravity, including the Sun and also Jupiter and Saturn, he predicted it would return in 1758. Newton died in 1727. Halley died in 1742. Science waited patiently. In 1758, after a year of suspense, the comet came around on Christmas Day, confirming the inverse square law of gravity and becoming the first object other than a planet known to orbit the Sun.

Apparitions of great comets are found throughout historical and prehistorical records. Observations in ancient Asia and particularly China were especially rich and go back thousands of years. The Mawangdui Silk from 300 BC includes illustrations of dozens of cometary forms, some of which resemble petrographic symbols, and identifies the disaster that each historical comet portended. It summarizes records going back a thousand years prior. Halley's comet was recorded as early as 12 BC, but there is no mention of its periodicity.

The second periodic comet to be discovered was 2P/Encke, an object that is nearing its finale and could be a future hazard to the Earth. It dances perilously among the inner planets and will either break apart in space or suffer tidal disruption near the Sun or collide with a planet. What we see today is a fragment of a substantially larger parent body. Its eccentric orbit takes it from inside Mercury almost (but not quite) out to Jupiter and back, every 3.3 years, so it has regular encounters with all the terrestrial planets. It is losing large volumes of material, and sand- to gravel-sized pieces plus some larger chunks are spread out along the orbital path. The Taurid meteor shower happens when the Earth crosses this trail of crumbs.

Shortly after Pan-STARRS, one of a new generation of survey telescopes, came online, the first interstellar interloper was detected. Named 1I/'Oumuamua, Hawaiian for *scout* (*I* for *interstellar,* the first such usage), it zoomed within a quarter of an AU from the Sun, with an interstellar velocity of 26 kilometers per second—way too fast to be gravitationally bound to our solar system. That's almost 6 AU per year. It came from beyond. Long and skinny, shaped unlike any comet or asteroid we've ever seen, 'Oumuamua was ejected from its birth star at least 300,000 years ago—that's the shortest direct flight from any of the closest candidate stars at that velocity, in the direction of Lyra, around 25 light-years away.

Is 'Oumuamua a natural body? I think so. Its color is dark red, like a primitive asteroid or comet. Plenty of comets are cast out from other solar systems that fall into ours, and comets are bizarre, and interstellar comets may well be stranger. When giant planets accrete, they toss out a large fraction of the objects that come near them. These become the cloud of comets around the star—their Kuiper Belts and Oort Clouds—but a fraction keep going, escaping the star's gravity entirely. Those would number in the trillions. Later, when a sunlike star dies and becomes a white dwarf with half its original mass, it releases most of its Oort Cloud to the gravity of the galaxy, another trillion comets to be mixed up with all the other stars. Lost comets are out there, yet space is big: if 1I is a random interloper, then it's been estimated that space must be teeming with about one Mars-mass worth of 100-meter asteroids per cubic light-year. That sounds about right, compared with how many solar systems are forming and how lossy the process is.

That said, 'Oumuamua is such a bizarre object that speculations abound, even among reputable scientists, that it could be a battered spaceship, or a fragment thereof. Like with so many other astronomical bodies, observations of 1I are limited to *photometry*, the analysis of a single faint pixel of its light. We can only estimate its size and shape based on how its brightness and color varied with time. Models are made of a dark object rotating in space, and least squares is used to derive the shape and tune the color and albedo[17] that best match the photometry. There are multiple solutions. 'Oumuamua could be shaped like a flying saucer, or it could resemble a submarine as usually depicted—either one would be unlike anything we've seen.

An alien spacecraft, you say, that dropped out of warp drive just in time (hence the slight wobble to its rotation!) and is disguised as an unremarkable rogue asteroid while it conducts a detailed reconnaissance of our solar system? It has a certain economy of explanation and might be a viable hypothesis if no more interstellar interlopers are discovered now that we're looking for them. If there is no 2I in the coming years, we will actually have to consider this more seriously. But from what we know, 1I is a natural, if strange and unusually shaped, comet-like body, and we'll likely see another before the year is out. Perhaps this is the same anticipation Halley felt.

Even if it's just a random comet, from the point of view of understanding alien life 'Oumuamua is remarkable. If fragments this large are going from solar system to solar system, big enough in principle to shelter rugged microbes, viruses, and spores for hundreds of thousands of years, then panspermia extends way beyond life between planets, to the idea of life crossing space and time using asteroids and lost comets as ferryboats: *intergalactic panspermia*. This doesn't have to succeed more than a tiny fraction of the time to alter everything.

GIANT IMPACTS OCCUR when full-grown planets end up on intersecting orbits. The accretion of a few dozen oligarchs, the standard model, would require about a hundred giant impacts. We know it ended about a hundred million years ago, the formation of the Moon being one of the last hurrahs. Because it takes about a million years for large debris to clear out of the inner solar system following a giant impact, terrestrial space was always crowded with innumerable scattered objects, many larger than Vesta and Ceres, diverse remnants from disrupted mantles, cores, and crusts. These would impact other planets, and cause gargantuan resurfacing collisions. The *late stage* of planet formation, when oligarchs get mixed and matched, transitions to the *late veneer,* when remnants get draped on top.

In a busy young planetary system, a near miss might excite two planets into crossing orbits and ultimately result in a giant impact. The new dynamical state—two bodies becoming one, plus debris—triggers further perturbations and collisions. Depending on how delicately it was set up, a complex system of planets can undergo a *dynamical catastrophe* and run amok with collisions until it finds the next stable state. That's the scenario Andreas Reufer and I have in mind for the late origin of the Saturn system described above.

It's also been proposed that the terrestrial planets are a second generation of planets: before this, perhaps for only a million years, there was a prior system resembling the majority of planetary systems that have been discovered so far. Most of them have Neptunes and super Earths, which appear to be the most common planets, and are more tightly packed than ours, and closer in. If a prior system interior to Jupiter was destabilized somehow,[18] then after the first collision occurred, it may have been a clash of Titans in an age of giant gods, or a line of semitrucks fishtailing in a high wind. If the collisions happened when the massive gas-rich disk was still around, then the largest planets would have been dragged into the Sun, consistent with the relative void that exists inside the orbit of Venus: strange little Mercury and nothing else. A guess as to the time scales might be a million years for the prior system to collapse and ten million years for the remaining scraps to duke it out, leaving behind a solar system such as ours emerging from its chrysalis or like a phoenix from the ashes.

It is speculations such as these, unfettered by any data that we know of and relying on models and conjectures, that cause us to look more closely at the Moon, a body we *know* was produced in a giant impact and that is the only reliable witness to what came next.

"All changed, changed utterly: / A terrible beauty is born."[19] Depiction of the standard model about ten minutes into the impact origin of the Moon. Theia came from the upper left at about 15 kilometers per second. The shock wave is seen as a circle propagating into the Earth and has almost reached the back of Theia, causing heating and melting. Converted to gray scale from original color.

Art © Don Davis

WHEN YOU ENVISION a giant impact, you have to slow down by a factor of a million. It's not a missile hitting a target; it's more like watching the *Hindenburg* disaster unfold. Although the collision speed is hypersonic (typically 10 or more kilometers per second) it would still take more than an hour for two planets between 3,000 and 10,000 kilometers in diameter to finish colliding. A giant impact is more of an attempt by two planets to pass through each other, and if they can't, for their cores to become mangled together and for the resulting larger object to grab what it can, representing a parameter called *accretion efficiency,* which is 1 when the bodies perfectly merge, 0.5 if the larger captures half of the smaller, and 0 in a hit-and-run, where no net mass is added (although mass can exchange). So it's a parameter that describes the outcome of a giant impact.

The process actually begins hours before the planets whack together. They travel toward each other at a few diameters per hour until they look like a binary planet, Pluto and Charon. They come under the spell of each other's gravity. As the final hours go by—at the speed of a lunar eclipse, if you want a comparison—the smaller one is deformed, so that by the time of the giant impact it resembles a rugby ball that has been spun up by the resulting torque. The velocity increases to the escape velocity plus whatever incoming velocity it had, so typical ramming speed for terrestrial planets is about 1.1 to 1.2 times the escape velocity—swinging around hard, as they say.

And then they physically connect. The process is disruptive and eruptive, like a gargantuan landslide spanning thousands of kilometers, but involving the deep mantles and the igneous cores of two planets shearing against each other. It plays out over hours or even days. Giant impacts have a lot in common with the formation of huge craters like Chicxulub and Imbrium, but they are global. Rock physics plays less of a role and astrophysics more, yet the

The yin and yang symbol is an icon from ancient China that reminds me of how two accreting planets swallow each other's cores.

event sets the stage for all of the geology that follows.

There are many variants on this theme of giant impact, so one shouldn't grow too attached to any one scenario. Still, it's useful to reference the standard model, where a Mars-sized body impacts the Earth at just above the escape velocity, about 10 to 12 kilometers per second, in a nominal 45-degree collision. The front line of the collision explodes in a violent arc as Theia and the Earth connect at ten times the speed of a rifle bullet. As the giant surfaces contact each other, it's hard to remember which is Theia and which is the Earth, which is up and which is down. Masses of mantle, crust, and ocean are jetted out of the interface like so much water from a nozzle, escaping to make weird planetoids and sprays of dust and vapor.

The collapse of crust and mantle and ocean and anything in between proceeds, and the smashed surfaces become sandwiched into the combined interior and then pulled apart like so much taffy. The cores, being dense, fall through each other's rocky mantle, sinking to the center of the collision where they become intimately connected, each swallowing the other. When it all settles down in a day or two, the deep mantle material is on top of the core, then other rocks, then the hydrated silicate minerals, then the ocean and atmosphere—a completely rebuilt planet with an uncertain memory of where it came from.

Two merging planets bring with them angular momentum, so the same principle that causes a protosolar nebula to become a flattened disk causes spiral arms to be spun out of the colliding materials. Indeed, a post-giant-impact structure can resemble a miniature galaxy with a central condensation and spiral arms with clumps. Depending on the kind of giant impact, the result can be a protolunar disk composed mostly of shock-melted silicate material around the merging planet—a "flying magma ocean," to use the phrase coined by planetary physicist David Stevenson, who first studied the thermodynamics of giant impacts in detail.

To understand some of the weird physics that ensue, go back to the day before the collision. Imagine a region in the mid-mantle of Theia, the Mars-sized planet that contains most of the stuff that will soon become the Moon. Soon it will be shredded into a disk around the Earth, but right now it is under extremely high pressure inside of Theia, just like the mid-mantle of any other terrestrial planet. Indeed, if you took a pound of Theia before the collision and teleported it onto the desk in front of you, it would explode with the same energy as a pound of TNT, just releasing from its pressure like a jack-in-the-box when you have just opened the lid. As the mantle of Theia gets shredded into a disk, this *enthalpy* contributes to the melting, vaporization, and expansion of the material that will become the Moon.

Returning to the idea of the proposed prior system of planets, of giant impacts involving super Earths and Neptunes, one thing is for sure: these collisions involve pressures and energies that are more intense, in proportion to more than the square of their size. If we scale to giant impacts involving tens of Earth masses, the explosive energy is so great that everything that isn't gravitationally bound into a massive clump—a remnant planet—is atomized. You might efficiently lose the hydrogen and other easily rarified elements, and from the recombination of "metals" into oxides, you might start to build the Earth and Venus. In all likelihood nothing would survive from before; this would be time zero for the rocky planets.

The hypothesis of a prior planetary system passes one sanity check: the total mass of the terrestrial planets, compared to Neptune, is about the mass of the Moon compared to Theia. That is to say, Earth and Venus are about the right amount of material we naively expect from giant impact scraps, if Neptune-mass bodies accreted, leaving behind their remainders before they migrated into the Sun. It's an exotic idea, but it is ultimately testable, because it probably requires that Venus, Mercury, Earth, and the Moon are of the same isotopic composition. (Conveniently for the hypothesis, we have no samples from Mercury or Venus.) If it is true, this would mean that our terrestrial planetary system isn't normal. The Earth and other planets would be the result of a highly selective process of attrition in which only a fraction of a percent of the starting matter survived.

CHAPTER 6

THE LAST
ONES
STANDING

Astronomy is nothing more than the outcome of
conjecture.

—SHEN KUO, *DREAM POOL ESSAYS* (1088)[1]

D ISAPPEARING ACTS ARE A POPULAR and entertaining
class of theories. First you invoke an object that was very
important and can explain everything. Then you make it go
away! Or you hide it. One popular hypothesis a century ago was
that a wayward star passed so close to the Sun that it ripped the
solar system out of its womb.[2] Of course you would never find the
star that did it, out there in the galaxy.

Sometimes the culprit is close enough to do what it needs to
do, but too far away to be detected just yet. These predictions can
be spectacularly successful, or not. As we saw, Ceres was predicted based on the geometrical progression known as Bode's law.
A massive Planet X was predicted to exist in the 1840s, based on
irregularities that had accumulated in the orbit of Uranus since its

discovery in 1781. Predictions were made for its location in 1846[3] using gravitational dynamics to calculate the gentle pulling by a hypothetical giant planet farther out. The same night they got the predictions, astronomers at Berlin Observatory pointed their telescope and beheld Neptune, the ice giant that frames the outer solar system, almost as if conjured out of ideas.

Other times the culprit never was. Supposedly hidden by the brightness of the Sun, the planet Vulcan was once invoked to explain the non-Newtonian precession of Mercury's orbit, which was eventually explained by Einstein's gravity. Today there is another Planet X—or a couple of them actually, one of them supposedly ten Earth masses and hundreds of AU away—close enough to solve the riddle of the inclination of the planets' orbits, but a thousand times fainter than Pluto, so we can be excused for not having found it. It will be discovered soon if it exists, once the Large Synoptic Survey Telescope[4] begins to pump out tens of terabytes of sky images every night. If not, it's on to Planet Y.

And then there's the idea of an original Saturn system that's hiding inside of Titan. How would you know? Or that a system of super Earths and Neptunes disappeared into the Sun, leaving Mercury, Venus, and the Earth. To study these scenarios, we have to look back into the chaos, and the pathway to the start may have been lost. I'm entertained by the idea of a prior solar system, but there's not really any evidence to support it. It emerged when it became apparent that most planetary systems are more tightly packed than ours, and that our system is anomalous. In another decade we'll have seen enough solar systems in detail to be able to think more clearly on it.

IN THE STANDARD model of Moon formation, Theia is like the midwife who delivers the child and is gone. Or is she still around?

According to the model, most of Theia came to rest inside the Earth—perhaps that could explain some of the major heterogeneities in mantle composition. But a significant remnant of Theia ended up in the sky. According to giant impact simulations, most of the Moon derives from the projectile.[5] That makes intuitive sense, that when the Earth accreted the core and deep mantle of Theia, the immediate result was a double spheroid that flung off the extremities of the interloper—the way you might lose your bag when jumping onto a moving streetcar.[6]

But while Theia may be hiding physically, and we have some understanding of that, she seems to have vanished chemically as well. Oxygen makes up almost 30 percent of the mass of the Earth, and 45 percent of lunar and terrestrial crustal rocks. The isotopes of oxygen ^{16}O, ^{17}O, and ^{18}O and other species like titanium, zirconium, and potassium are like flavors whose ratios tell about a rock's country of origin. Their atomic behavior is basically the same, but the masses (numbers of neutrons) are different, so they serve as labels. The systematic ratios of oxygen isotopes for Earth rocks are the same, as expected if Earth started from a common or well-mixed reservoir. They are isotopically distinct from Mars rocks, and both are significantly different from the ratios of oxygen in the Sun. Meteorites also have diverse isotopic ratios, but it's confused by the fact that we don't know what makes the reservoirs distinct, and there's a lot of uncertainty in the data.

Lunar oxygen is indistinguishable from Earth oxygen to ten parts per million. Isotopes of titanium, an element that does not behave chemically like oxygen, are similar to within four parts per million. Zirconium is also similar, and so on. Moon rocks and Earth rocks appear to a high degree of confidence to have derived from the same isotopic reservoir—and yet there are differences. The Moon is depleted in ^{39}K (potassium) relative to ^{41}K, but this is consistent with the idea that lighter isotopes are more easily vaporized and will be lost, like water was, in the intensely shocked post-impact state, showing up especially in these "semi-volatile" elements.[7]

By the late 1990s, although the geochemists were finding evidence to the contrary, the giant impact theory was taking shape as a convincing dynamical model. It explains the lack of metallic iron in the Moon, the large angular momentum of the Earth-Moon system, and the low H_2O abundance of Apollo samples. The shock heating would produce a lunar magma ocean, which is the required starting condition for the anorthosite lunar crust. To top it off, it explained these things by invoking an event that was (soon enough) acknowledged to be intrinsic to late-stage terrestrial planet formation, the mergers of oligarchic planets.

Taking a completely different approach to the question of planet formation, scientists were at the same time patiently measuring isotopic ratios of minerals from Apollo samples, using ever more impressive techniques. Their results would soon tear the original giant impact theory from its hinges.[8] The most basic clash of facts—that the Moon is made mostly of Theia, but there are no obvious geochemical signs of Theia in lunar samples—has led to a very creative twenty years, in which ideas for lunar formation[9] have drifted around like so many planets, sometimes colliding and accreting. But otherwise it's how British astrogeophysicist Harold Jeffreys described his field[10] back in 1929, that our diverse theories have become like "a lumber-room full of untested hypotheses [in need of] an occasional spring-cleaning and bonfire."

Maybe Theia hit the Earth with so much energy that the whole thing exploded, starting over as a homogenous mixture.[11] Maybe Earth and the protolunar disk somehow exchanged nearly all their oxygen after the giant impact. Maybe the Moon *is* of distinct composition, but later on was buried under hundreds of kilometers of crust that came from the Earth. Maybe there were many smaller giant impacts. Maybe Theia derived from the same isotopic reservoir as Earth—a premise that would solve everything, but has profound requirements.

George Darwin's theory from 1879 would have handily solved the crisis, had the physics worked out. In his theory, briefly described in chapter 1, the Earth spun so fast that the Moon popped out of the mantle—the Pacific basin, some would later say. He first developed the theory of tides, whose action pulls the Moon away from the Earth. It was much closer in the past, and the rate of tidal migration was therefore much faster. Every action (expanding the Moon's orbit) has an equal and opposite reaction (slowing the Earth's spin), so if you wind it back to the beginning and assume all the matter is merged into one planet at the center, you're talking about a pre-Moon-formation Earth spinning with a period of 5 hours.

A planet spinning that fast would be somewhat out of round, like asteroid Vesta (period 5.3 hours). But to fling off a satellite, a planet needs to be spinning more than twice as fast. Knowing nothing of giant impacts or of Mars-sized planets that had come unmoored, Darwin proposed that the Sun was able to add energy to the tidal bulge through a resonant action until the bulge crested and the Moon erupted, like Athena springing from Zeus's forehead.[12] Yet even if the detachment were to work out, this moon-blob would probably fall right back or escape; getting caught in a near-circular orbit would require precise conditions. And then, even if a large blob ended up in orbit, it would be orbiting so fast (every 2 hours) that its own tidal bulge, lagging behind it, would soon drag it down. His moon needed an extra big kick.

We can sit here and poke holes in Darwin's theory all day, and that's the point. His was the first scientifically consistent and therefore falsifiable model for the origin of the Moon, and we keep coming back to it. The fact that his starting scenario doesn't work is irrelevant; his tidal modeling set the stage for what followed.

And the end state of a giant impact accretion is geophysically and dynamically compatible with what Darwin's theory requires. It deposits too much angular momentum in one place, and causes the Moon to pop out of the mantle . . . but the wrong mantle, it would appear!

Any lunar origins model has to create a Moon that orbits at least a few Earth radii away, because otherwise it gets dragged back down. This can explain why Mars has only its two little moons, potato-shaped bodies Phobos and Deimos, orbiting at 3 and 7 Mars radii, measuring 22 and 12 kilometers diameter respectively. Mars has no sizable moons not because it is small but because it rotates too slowly to keep a massive moon from falling back down.

When a debris disk forms around a planet following a giant impact merger, there is a zone inside the *Roche limit*[13] where the planet's tides will shear a satellite apart before it can accrete. For rocky planets this corresponds to an orbital distance of about 2.5 radii and an orbital period of about 8 hours. So let's say there is a giant impact that makes a debris disk, and that a massive moon begins to form right outside the Roche limit with an orbital period of 10 hours. For the early Earth's spin period of 5 hours, a newborn Moon beginning at that distance would be spun up steadily into higher orbits. Within 10 million years it would be tens of Earth radii away, well on its way to where it is.

Satellite condensation beyond the Roche limit, in the case of Earth, places the Moon on an outward-bound spiral. That is not the case for Mars. The reason Mars can't make a big moon is that its day, 25 hours,[14] is five times slower than the early Earth's was, and probably hasn't changed much. (Mars has no major satellite that could have slowed it down, the way the Moon slowed down the Earth.[15]) It's been proposed that the formation of Borealis Basin created a protolunar disk that, per the simulation, accreted just outside the Roche limit[16] to make a new moon of Mars. The

trouble is, this new moon would be orbiting inside the danger zone, faster than Mars rotates. It would spiral in instead of out, and soon crash into the Martian surface—a fate that could take a few years or a few thousand years depending on the mass of the moon and where it orbits, and the tidal friction it creates[17] and the effect of an atmosphere.

Phobos at 2.8 Mars radii is well inside the co-rotation radius, so is spiraling in, and is predicted to crash in 40 million years.[18] One wonders that we should be so lucky to see it at all. When it collapses, it will be the most spectacular event and leave a new little Saturn in the sky. Phobos will break up as a debris ring when it gets much deeper

An image of 22-kilometer diameter Martian satellite Phobos by the High Resolution Stereo Camera onboard the Mars Express mission.
ESA/DLR/FU Berlin

inside the Roche limit. The debris disk will spread, spiraling down onto the planet, kicking up dust and making a band of craters. Thinking further on the theme of disappearing acts, it has been proposed[19] that Mars *did* have a massive original satellite, thousands of times more massive than Phobos and Deimos—the size of Vesta—that formed in the giant impact but then crashed right back onto Mars. Its short life would not be in vain. During the death spiral it would interact gravitationally with any outer moonlets, flinging them into long-lived higher orbits—perhaps even a repeated cycle of descending lunar breakups.

If Phobos and Deimos are just scraps of an original much larger satellite of Mars, then it would have crashed onto the Martian surface at 3 to 4 kilometers per second as an equatorial band.[20] Geologically this band or layer would be well disguised as sediments and pyroclastic (airborne igneous) deposits, with some mixture of satellite materials, perhaps the earliest geologic record on Mars after the formation of Borealis. It might no longer be evident given the relatively vigorous cycles of erosion and deposition and volcanism that Mars has experienced since.[21]

Earth and Venus themselves represent a major disappearing act. Growing to twice the diameter of Mars, they acquired 93 percent of the rocky matter that survives between the Sun and Jupiter. They each accreted almost a dozen Mars-sized embryos, which in turn each grew by accreting tens of Moon-sized planetesimals, and so on—a *feeding chain* of planetary growth, most of the bodies disappearing inside of larger bodies. Imagine, for a moment, that proto-Earth and proto-Venus were the two largest sharks in the ocean of the inner solar system, and that they fed well on almost all the smaller sharks. This is the standard view, but it has a specific nuance related to *attrition* that is overlooked, one I consider crucial.

If Earth and Venus accumulated *all* the smaller sharks, then no matter whether smaller sharks ate still smaller sharks, there would be no record of how the process played out. The record would have disappeared, whether there was a feeding frenzy among the small sharks, leaving three or four next-largest sharks, which then got eaten by Earth and Venus; or whether Earth and Venus gobbled all the small ones directly. It wouldn't matter, and you wouldn't know.

But planetary accretion is not even approximately efficient; one of the defining characteristics is that sometimes the smaller sharks get away. What if, for every nine smaller sharks, proto-Earth and Venus missed one? And what if these smaller sharks in turn missed a small fraction of the next-smallest sharks? Now you would end up with some very hard to catch next-largest sharks, and a lot of even harder to catch next-next-largest sharks, and so on, plus the "winners" of accretion, Earth and Venus.

Observing the population after planet formation was finished, you might be forgiven for thinking that those wily, lucky sharks are representative of the beginning population, but you would be mistaken. They are not random, but were selected for having evaded being caught.[22] Here's another example of what we call attrition bias: Consider a hundred soldiers who head out, unremarkable young recruits with limited experience, coming from common backgrounds. (These would represent the embryonic planets from here and there in the inner solar system, the starting bodies of planet formation.) Each soldier has his or her own unique qualities that are about to be tested and his or her own good or bad luck. After a brutal campaign, let's say only ten made it back, and the rest were buried. Those who returned had extraordinary qualities and combat skills, and extraordinary stories of good fortune. Maybe one or two were deserters who avoided the clash of armies. The earth accreted them with every burial, and as a consequence of their attrition, what's left is a highly diverse population of fabled veterans.

BORN AMIDST CHAOS and prone to instability, planetary systems start off as unpredictable as spring weather. They evolve over millions of years into a stable state where planets and their moons have learned by trial and error to avoid one another, either by staying away or by finding resonant orbits that don't collide. Kind of like people.

The resonances of Jupiter's Galilean satellites prevent them from moving independently from one another, resulting in the orbital stability of Io, Europa, and Ganymede. Likewise, Neptune never collides with Pluto. Tethys never collides with Calypso and Telesto, even though all three share the same orbit around Saturn. Other resonances are more subtle. For example, Venus orbits the Sun almost exactly thirteen times for every eight times Earth does, tracing out a five-point flower pattern. This indicates that Earth and Venus may have been connected in some fundamental way during their formation.

Venus and Earth ended up with about the same fraction of iron and rock, and grew to be about the same size, so every connection and distinction is important. Venus rotates the slowest of all the planets (a period of 243 Earth days, retrograde) and has no satellite, but I'd say its similarities with Earth outweigh its differences in the big picture. In the hypothesis of a prior solar system, Venus and Earth are wily sharks who avoided being eaten by super Earths and Neptunes by piloting the chaos. If so, my feeling is that they would have ended up more different than they are, like Mercury and Mars. Instead they appear to have been sharing an ocean together, the largest sharks, growing up in slightly different ways.

If the Earth is at rest in the center of this "Venus Rose" diagram, then Venus traces the rose pattern like a Spirograph, sometimes coming closer and sometimes farther, as both planets orbit the Sun. The apparent movement of Venus makes this splendid five-pointed shape because Venus orbits the Sun very nearly thirteen times for every eight times that the Earth does, for reasons we do not know.

Another paradox, that of the "warm, wet Mars," could be attributed to planetary dynamical chaos. Half a billion years after the planets formed, when the Sun was putting out only three-quarters of its present luminosity, Mars experienced atmospheric and climate conditions that gave rise to meandering river channels, catastrophic floods, and chains of crater lakes. Yet Mars gets only 43 percent as much solar energy as the Earth, per square meter, and it would have been even colder if the Sun was putting out three-fourths as much heat. To create warm, wet conditions, according to climate models[23] Mars would need a 2-bar CO_2 atmosphere to hope to have a surface temperature at the liquid point of water. Then, after doing its job as a greenhouse gas, the gigatons of CO_2 would have to disappear without a trace. The missing carbon would be evident in the rock record, the way that banded iron formations from 2 billion to 2.5 billion years ago coincided with the sudden rise of oxygen on Earth, which is worthy of a brief digression.

Photosynthesis began on Earth over 3 billion years ago, producing oxygen—specifically O_2, the free oxygen found in the atmosphere. Most life on Earth was unaccustomed to this reactive poisonous gas, but that was okay because the O_2 was removed as fast as it was produced. It oxidized the rocks, making them red (e.g., rust, FeO). But around 2.7 billion to 2.4 billion years ago, mats

and blooms of photosynthesizing *cyanobacteria* went wild and pro-
liferated in the waters and on land, causing the Great Oxygenation
Event that ended the Archean and began the Proterozoic, which
would ultimately lead to the rise of complex life. So we'd like to
look for these sorts of major records on early Mars, and the collapse
of almost the entire atmosphere would be one of them.

If Mars had a massive CO_2 atmosphere and abundant surface
water (the "warm, wet" scenario), then CO_2 would have dissolved
in that water and would be precipitated as carbonates. If 2 bars of
CO_2 were lost that way, there should be meters of carbonates all
over Mars, a distinctive mineral. There aren't. Although there are
some carbonate outcrops, for the most part what's found is a trace
that could be accounted for under present climate conditions. Another idea is that CO_2 was lost to the solar wind, since Mars has
a weak magnetic field and a low escape velocity. But if the solar
wind was so intense as to remove 2 bars of CO_2 from Mars, you'd
think Venus would have lost most of its atmosphere as well, being
exposed to 5 times the solar intensity and without a magnetic field.
Without a clear explanation for the warm, wet early Mars, maybe
it's worth considering an alternative explanation—which brings us
back to planetary chaos.

Mars lives in the dynamical shadow of the giant planets. In the
original Nice model, where Jupiter and Saturn migrated through a
2:1 orbital resonance 3.9 billion years ago, the terrestrial planets
were left on excited orbits, especially Mars; this was interpreted as
a problem for the model, leading to the "jumping Jupiters" scenario
described above. But on the other hand, a highly excited orbit of
Mars could be an excellent explanation for past evidence of flowing water, even in the absence of a massive atmosphere, so long as
Mars ends up on its present, less-excited orbit.

Mars today has an orbital eccentricity $e = 0.1$; in other words, its
perihelion is at 1.4 AU and its aphelion is at 1.7 AU. Solar heating

is 45 percent more intense at perihelion than aphelion, contributing to a complex seasonal cycle. What if Mars had been provoked in the past to even higher eccentricity, like $e = 0.3$? Then it would swing from 1.1 AU, experiencing almost Earthlike sunshine for about six months, then out to 1.9 AU for a deep, extremely hard winter for about fifteen months. This freezing, then thawing, would be a recipe for powerful hydrologic cycles that would thaw the permafrost, melt the ice caps, and cause catastrophic floods to pour into the northern lowlands, for as long as the carnival lasted.

Is it a crazy idea? Probably—you'd have to explain how it got back to behaving like a "normal" planet with eccentricity of 0.1. But no more crazy than a 2-bar atmosphere that disappeared without a trace. And it was only a slightly more wayward Mars equivalent that made the Moon according to the standard giant impact theory, so you be the judge, whether Mars was lucky and Theia was not.

TODAY THE ORBITS of the major planets are stable on time scales of billions of years. Over the past 2 billion or 3 billion years, only relatively small collisions have occurred—for instance, the one that killed off the dinosaurs. Kilometer diameter asteroids hit the Earth every million years, forming impact craters and oceanic cavities. Sometimes an asteroid or comet breaks up in the inner solar system, causing a torrent of smaller impacts for some time. Hundred-meter near-Earth objects (NEOs) impact every 30,000 years, give or take, most of them making several-kilometer holes in the ocean and the rest of them going undiscovered, buried in jungles or filled in by sediments.

We are, for obvious reasons, interested in where and when the next geologically significant crater will form, but for that, unfortunately, we have only a statistical answer, because the location of an object that has regular close encounters with the Earth and

Moon is chaotic on time scales longer than around three hundred years, and nothing is on the known horizon. Think about it: no *known* asteroid has a greater probability of hitting Earth than the probability that a *random* asteroid of the same diameter will hit us first.[24] That's about as good as it gets.

Today there is a lot of activity exploring and understanding NEOs, but fifty years ago few appreciated them or acknowledged the impact hazard. The holes in the Moon were believed to have formed eons ago. Astronomers, many of them hobbyists, collected asteroids like stamps, as curiosities that could be named. Only a few studied them as a research focus. Comets were more exciting, exotic novelties with ever-changing activity that were yielding fantastic data from spectroscopy during their bright apparitions. Only a few impact craters were recognized on Earth, including Meteor Crater and a few large ones like Popigai in Russia.[25] More to the point, until the 1990s we had never seen a picture of an asteroid— they were just dots in the sky, befitting their name, meaning "star-like."

The mainstream scientific appreciation of asteroids came from an unexpected direction: sedimentology, and this takes us back to the Cretaceous. The Earth is draped with 20,000 tons of cosmic sediment per year, mostly in the form of dust- to gravel-sized meteors impacting the upper atmosphere. Boulder-sized meteors make it deeper, exploding 30 to 50 miles up, producing dust and small fragments (meteorites). Interplanetary dust particles (IDPs) are stopped before there is much heating, so they drift down to Earth somewhat intact[26] and contribute to the cosmic pollution on your roof.[27]

Cosmic dust becomes part of the sediment that piles up at the bottom of the sea. It is a tiny fraction compared to the terrestrial sediment that is discharged by the muddy rivers of the Earth. When continents are rising and it is raining a lot, there is more sediment discharged, and cosmic dust will be diluted still further. If the climate is cold and dry, on the other hand, there is little sedimentation, and the cosmic dust component will be high. So if you can measure the cosmic fraction of dust in a fossilized seabed, you'll know whether the erosion rate on land was fast or slow, and thereby the vigor of the hydrologic cycle, of clouds and rain.

Cosmic dust is primitive and has elements that are strongly depleted in the Earth's crust. One of these that is relatively easy to measure is iridium, a metal whose abundance in the crust is merely two parts per billion. Iridium and other *siderophiles* (metal-loving elements like gold and tungsten) vanished when the Earth melted and differentiated, catching a ride with the iron into the core. IDPs are mostly primitive undifferentiated material,[28] so their iridium is thousands of times greater than in Earth's crustal rocks, so that while you might not see the cosmic dust with your eyes, you can see the excess iridium using a mass spectrometer. If the rate of cosmic sedimentation is 20,000 tons per year over the surface of the Earth, then the amount of iridium in a layer is proportional to the passing of time.

The Mesozoic-Cenozoic mass extinction, also known as the K/T or the end-Cretaceous,[29] was the exception to the rule. This boundary of major geologic epochs 65.5 million years ago is expressed as a thin clay layer in seafloor sediments worldwide, and sometimes near-surface deposits[30] that were buried before they could be eroded. There is a strong iridium anomaly in this boundary clay layer, indicating a lot of cosmic dust, meaning that it was formed over a long time, and during a period of slow erosion. The cosmic sediments would have to build up for millions of years. Not only had the dinosaurs died out, but there had been a global resetting of all kinds of life-forms, from the Triceratops down to the diatoms and planktons in the sea.

To have such a strong iridium anomaly, the rivers on Earth must have slowed down to a trickle. A whole geologic period would be sandwiched inside this thin layer—a dry or frozen planet where the continents and oceans were covered in ice for millions of years, a snowball earth.[31] Or maybe Earth had become a desert planet where it did not rain for millions of years.[32] Such explanations did not make sense because the Cretaceous was a peak time for dinosaurs, featuring swamp creatures and jungle eaters who for the most part thrived, and oceanic flora; it wasn't an ice age or a Mars-like desert planet, and there was no hint that it was about to become one. So in the end the K/T boundary went unexplained, leading to great confusion about how the dinosaurs went extinct.

One of my early acquisitions was a View-Master, a popular 1960s toy that you see no more, what with everyone gazing into their phones. I was given it on a family vacation through southern Utah, plus a half dozen image disks that had stereo pairs, one for each eye. You would hold the View-Master up to the light like a pair of binoculars and advance the lever from 3D picture to picture. The T. rex really popped out! So did the Stegosaurus, whose weaponized tail would defend it from the bluish monster. In the background was a canyon desert, like Utah, and for good measure, volcanoes were going off—it was a terrible time! But there were also palm trees. The landscape didn't make any sense to me at the time, but now I get it: nobody had any idea.

What if you were to question the premise, that iridium measures the infall of cosmic dust? In 1980, Berkeley geologist Walter Alvarez and his father, physicist Luis Alvarez, and their colleagues proposed that instead of a million-year buildup of dust, the K/T came as a single large "bolide" that delivered a million years' worth of dust in one day—a gigantic meteorite. Their paper was the sort of revolutionary analysis that comes about when people (including family!) from totally different disciplines and perspectives team up to reconcile a piece of incontrovertible data that has wrecked the prior understanding.[33] It was something of a Copernican moment, if you ask me.

The amount of iridium represented in the K/T layer, altogether, is equivalent to what's found in an asteroid 10 kilometers across. So this asteroid impacted and ejected a 100-kilometer hole from the ground, like a slow-motion cannonball landing in water. The splattered asteroid added on average about a 1-millimeter layer of meteorite, mixed with centimeters of dust-sized ejecta and materials scoured where the ejecta landed. Distant ejecta reimpacted at 5 to 10 kilometers per second, not only devastating the landscape but producing an incendiary reentry of fireballs that could have burned down forests and roasted animals by the heat.[34] The oceans were acidified, the forests burned, and the skies darkened with smoke for weeks—not much made it through it.

The ash layers, the survival of burrowing species (our forebears), the devastation of plankton in the acidified seas, the global shutdown of photosynthesis—these pieces made sense in the framework of this new scientific theory that the K/T layer did *not* form over millions of years. It formed in one day, and in the dark burning aftermath that followed. And as scientists went exploring, finding further pieces to the puzzle (submarine landslides induced by the impact; carbon isotope excursions), the story became increasingly rich. The clincher was the identification in the early 1990s of the crater itself, named after the fishing village Chicxulub near the measured epicenter. (The crater, being one of the largest on Earth, is rather shallow, and completely buried under Tertiary sediments.) Eventually fragments of chondrite meteorites were found inside the K/T boundary deposits, the bolide that did the deed.[35]

THE SMALLEST CRATERS are pockmarks governed by rock fracture; meteors this small only reach the surfaces of planets that lack atmospheres, like the Moon. (Many lunar samples have these pockmarks.) Then there are steep-walled holes that look like you could have dug them with a pickaxe, and then those the size of open-pit mines and volcanoes. Once planetary craters get too big, their floors slump flat and they become like broad, round valleys except without an entrance or exit, many miles across. Craters

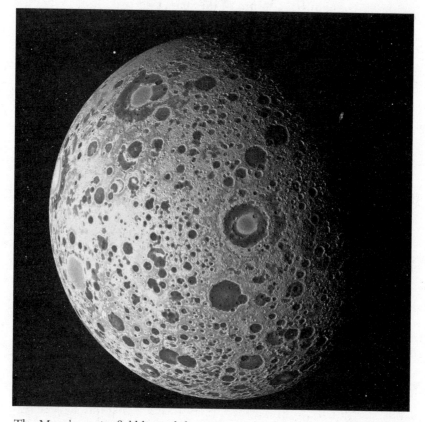

The Moon's gravity field has subtle variations that were measured by NASA's GRAIL mission in 2012. These gravity variations are associated with mascons (dense mantle plugs filling the large craters) and mass deficiencies (removal of material by impacts). The color image has been rendered in gray scale, so dark features can be high or low gravity, associated with large craters that make long-lived holes in the crust. The image is centered over Mare Moscoviense on the farside.
NASA/JPL-Caltech/MIT/GSFC

larger than about 2 to 3 kilometers diameter on the Earth, and 10 kilometers on the Moon (due to the lower gravity), collapse into shallow pie-tin shapes with complex crustal geology. Larger than that, they form *central peaks* where the crater collapses and rebounds. For large complex craters, the pulling in and fallback and uplift of material can cause regional responses and gigantic tremors, even changing the nature of the planet's crust.

Think of digging a hole to plant a tree. The first shovelfuls are easy: you're just moving some soil from here to there. But it gets much harder to dig, the deeper you dig—the soil is under more compaction, harder to penetrate, and you have to lift each shovelful higher and farther. It takes more energy to work against pressure and gravity. Now apply this principle to planetary scales of kilometers, and to hundreds of kilometers, and at some point, it takes so much energy to dig out the rocks that you transform the rocks, and the intensity of the excavation melts them. The bigger a planetary impact basin, the more energy is required per "shovelful," so the largest craters can melt the whole region in which they are forming, leaving no crater, only hints of it—a resurfaced planet.

The biggest crater on the Moon, if it is a crater, is Oceanus Procellarum, spanning more than a quarter of the Moon's circumference, almost 3,000 kilometers. It makes up most of the Man in the Moon and encompasses a smaller, less ambiguous impact structure, the 1,100-kilometer diameter Mare Imbrium, which stands out in gravity maps of the Moon. Although to an astronaut the gravity of the Moon feels rather constant, it actually changes from place to place. In the case of Imbrium and other major impact structures, dense mantle material flowed up to fill in the holes dug out of the lighter crustal rocks by large impact craters. Procellarum doesn't have a mascon per se—it is so big that the whole region reestablished gravitational equilibrium. But its perimeter is demarked by gravity gradients, like some kind of

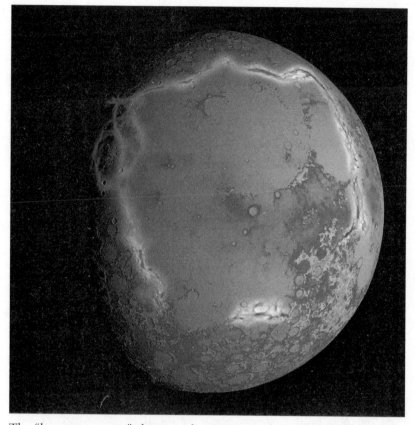

The "lunar pentagram" shows up by projecting the *gradient* of the gravity field onto the topography of the nearside around Procellarum. This measures the amount that the local gravity field changes, too subtle for an astronaut to feel but geologically significant. Rendering in gray scale makes the data values ambiguous; for an animated festival of color go to https://svs.gsfc.nasa.gov/4014. Combination of GRAIL (Gravity Recovery and Interior Laboratory[36]) data and Lunar Reconnaissance Orbiter data; image processing by Colorado School of Mines and NASA's Scientific Visualization Studio.
NASA/JPL-Caltech/MIT/GSFC/CSM

horse fencing around a big pasture. Unexplained patterns show a deep, abrupt (although slight) contrast in the density of the crust or upper mantle, what looks for all the world like a "lunar pentagram";

or if I were searching for geophysical analogies, I'd say it would resemble a mud crack.

It is not known whether Procellarum is an impact structure or several overlapping structures. (And what are the odds of that?) It is also not known whether the bounding lunar pentagram is related to an impact or came much later as part of a global adjustment, some kind of tectonic process that could be related to the lunar hemispheric structural dichotomy—a thick farside crust and a thin nearside.

Another major early impact on the Moon is the 2,500-kilometer diameter South Pole–Aitken basin[37] on the southern farside (the South Pole is on its southern rim; crater Aitken is on its north; it is unimaginatively named). The SPA is the deepest (13 kilometers) and probably the oldest surviving lunar crater. The region has been battered and buried under subsequent craters, so although the SPA has been recognized since the 1970s, its geology was nondescript until the Galileo spacecraft flew by the Earth-Moon system for one of the gravity assists that would slingshot it to Jupiter. It pointed its state-of-the-art spectrometers and cameras on the farside, the first data set of its kind, and revealed a compositionally distinct central region to the SPA, a crater that had punched through the thick highlands anorthosite.

We can estimate the diameter of the impactor that made the SPA and the giant nearside basins only so far as we can estimate the volume of the giant hole that existed immediately after the asteroid or comet struck. It would have collapsed right away, just like the Eltanin crater in the ocean, but not completely. The result is a shallow, wide basin, and after four billion years of further evolution, we have to guess what did it. A good estimate is that Procellarum and the SPA were made by projectiles a tenth the size of the Moon—major asteroids the size of Psyche or Vesta. Although not giant impacts, they are getting up there.

The physics of mega-cratering can be scaled up from common experience. Returning to our familiar comparison, when a bullet first impacts a block of wood, it crosses its own diameter in a microsecond. During a mega-cratering event, this interaction is slowed down by a factor of millions, proportional to the larger scale. It takes a second or two for a good-sized asteroid to puncture the lunar crust (divide the crustal thickness by the impact velocity). Something big enough to make Procellarum or SPA would require another twenty seconds to bury itself into the Moon. And then once this "contact and compression" phase is completed, the opening of the crater itself would take ten minutes. Although you would not want to miss it, you'd have time to enjoy a cup of coffee while the event played out.

THE IDEA THAT there were two moons, and that one of them became the farside highlands, began as science frequently does, by thinking about something quite different. Martin Jutzi of the University of Bern and I were studying the weird shapes of comets, developing models for collisional crushing and deformation so that we could find out how they accreted[38]—how fast and from what. Comets, it had been observed, are frequently "duck shaped," with two unequal lobes stuck together like 67P/ Churyumov-Gerasimenko. New Horizons target Ultima Thule is another example. Other times they have the appearance of layered piles, an idea advocated by Kitt Peak astronomer Mike Belton, a pioneer of comet geophysics who was also the imaging lead for the Galileo mission to the Jupiter system.

The early outer solar system is thought to have accreted in multicar pileups,[39] and Belton envisioned that the constituent material was soft and fluffy, like down pillows, so that they would form accretionary structures he called *talps* (*splat* spelled backward).

Cross your eyes. Stereo left-right pair of the duck-shaped Comet 67P/
Churyumov-Gerasimenko and its jets. If you make the images line up, then
your brain will combine the pair, separated by 1.2 degrees, to produce a stereo image. At the time these images were taken the 4-kilometer comet was
spewing dust and gas at a rate of tens of kilograms per second.
ESA/Rosetta/MPS

The measured bulk densities of comets are only half the density of
water ice—they must indeed be fluffy, easily deformable things.[40]
If you have built a snow fort, you will know the process: whack a
lump hard enough that it splats, contributing to the wall; then do
it again.

Certain comets have the look of layered piles, and others look like
rubber duckies. After the Deep Impact flyby[41] of Tempel 1, which
appears massively layered, Martin and I had begun simulating these
hypothetical collisions to find out what an impact would imply physically in terms of observables. If cometesimals accreted pairwise
at only a few meters per second, then the collision would happen
slowly, a scaled-down giant impact, taking hours. Although only
the speed of a bicycle crash, it would have a lot of momentum—a
mountain riding a bicycle—and this sort of collision was beyond
any laboratory experience and well outside our limited intuition.

To study these slow crashes, we had been improving and testing our code to incorporate porosity and crushing and friction—reproducing experiments of sand in a hopper, small landslides and so on. It turns out that including fragmentation and crushing and strength and friction is quite challenging computationally—it may seem surprising that modeling dirt requires supercomputers, but it does. We were discussing how to speed up the simulations;[42] eventually we needed a break.

On Fridays my department held a brown-bag seminar,[43] this time on the shape of the Moon. Everyone knew the basics—that the Moon is slightly oblong, aligned with the Earth's direction, and so forth. But it is much too oblong to be accounted for by today's tides. Also, according to gravity models, the Moon's crust, made out of rocks that are about 20 percent less dense than the mantle, is piled as thick on the farside as on the nearside, up to 70 kilometers. The seminar considered the theory[44] that the Moon's

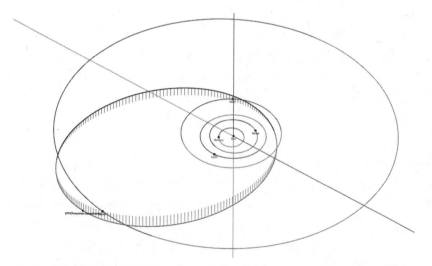

As I write this, Rosetta mission target 67P is at an interesting point in its orbit, slightly beyond Jupiter and closely trailing it, gaining a little bit of extra velocity like a skater pulling on a tow rope. It will swing by again in 2021, then 2027, and so forth. This is a typical JFC orbit, always having to mind the planets.
Image by the author, using online visualization tool provided by NASA/JPL/CNEOS at http://ssd.jpl.nasa.gov

oblong shape was frozen in about 4.4 billion years ago, when it was orbiting much closer to the Earth and spinning faster. Dynamics aside—it is difficult, perhaps impossible, to make this work—the measured topography of the farside bulge was found to be a good fit to the theoretical prediction. But, I wondered, wouldn't a spinning, tidally distorted Moon make bulges equally on both hemispheres? What happened to the lunar nearside?

Until 1959, there was no direct knowledge of the farside of the Moon.[45] The farside had been a place of pure imagination until humankind's best, first foray in the race to the Moon. The Russians started the space age in 1957 with Sputnik, and by 1959 they had proved, to the terror of many Americans, that they were capable of sending spacecraft to the Moon. In September 1959 they sent Luna 2 to crash into the nearside, the first human object sent to another planet. The massive (almost three hundred kilograms) and more scientifically motivated Luna 3 was launched a few weeks later, a classic product of fifties technology. It featured a film camera system adapted from spy planes, one that would take wide-angle and telephoto pictures. (Digital imaging was decades away; the United States would fly the first television camera in orbit the following year.) But the Russians were missing a key technology: a film that could survive the hard radiation and extreme temperature conditions of space without fogging.

It turns out such film had been developed by the U.S. military and was being used for spying on the Russians from high-altitude balloons. The United States would launch them into the jet stream from Europe, collect them in Alaska, and develop the film and analyze the images, along with other data. On the coldest Siberian mornings, these balloons would lose elevation to the point that they could be shot down by Russian MiGs, and engineers would pick apart the American technology. Among their finds were radiation- and temperature-tolerant film, much of it unexposed. Many

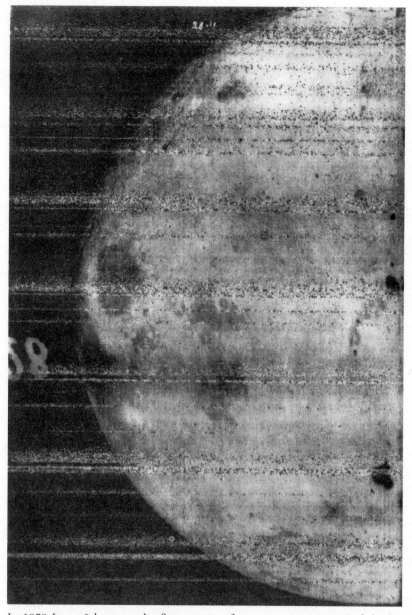

In 1959 Luna 3 became the first spacecraft to capture an image of the lunar farside, using spy-camera technology and purloined U.S. high-altitude surveillance film, as can be read in detail at http://www.svengrahn.pp.se. Although the image is noisy and low resolution, it is clear to see it is a wildly different landscape from the nearside.

Roscosmos/IKI

years later the chief engineer of the Luna 3 camera system revealed that he had secretly taken this American balloon (AB) film and trimmed it to size, and loaded it in the Moon-bound camera.

So the first images of the lunar farside were projected by a Russian camera onto U.S. spy film, processed chemically onboard the spacecraft, and rasterized by scanning a bright light source behind the fixed image onto a photomultiplier tube, turning it into an analogue radio signal that could be beamed to Earth in the weeks after the encounter. The film image itself, onboard the spacecraft, would have rivaled modern lunar images in quality, but due to the scanning process, it was returned as a low-quality fax. Film images would continue to be used in space for quite some time, in part because of the military application of similar technology; the five Lunar Orbiter missions in 1966–67 would use advanced film scanning to create some of the most exquisite pictures ever taken.[46]

The first image of the farside was a shock. Instead of dark maria and arcs of mountains, it was a gigantic plateau extending for thousands of kilometers, a battered shield. Although the Moon has no plate tectonics to make such a shield (not enough heat, too little gravity), the clustering of maria on the nearside and the thick crust on the farside does have a resemblance to the relatively recent (late Permian, 250 million years ago) history of Earth, when the supercontinent Pangaea was on one hemisphere and the Panthalassic Ocean on the other. Earth too has had its asymmetric moments. So maybe lunar mantle convection tore apart the nearside. Maybe a rain of impacts did it, by luck of the draw, removing one hemisphere. Maybe the Earth's tides did it.

Maybe another moon did it.

THE SEMINAR WAS drawing to a close and my mind began to drift. I thought about how big an impact would be required to obliterate the nearside of the Moon but not the farside—an impactor around 500 kilometers in diameter, the size of Vesta, perhaps. I thought about the probability of hitting the oblate Moon smack-dab on one of the bulges, and how the Earth would figure into this, 10 or so radii away when the Moon was solidifying. Maybe a bunch of trapped gases ended up beneath the nearside but not the farside, for some reason, and escaped through the volcanoes, causing the nearside to deflate like a punctured tire . . . I went down a rabbit hole of thoughts. I glanced toward Martin and made a gesture of my palm colliding with my fist, referring to the research we had been doing on planetesimal splats. He kind of shrugged, as if to say, *Sure, why not?*

We returned to his office to set up the initial conditions for a simulation that would make a big splat on the Moon. We knew it had to be a relatively low-velocity collision, because if it was at 10 kilometers per second, you'd just blow the Moon to pieces. In our comet models, we were colliding things at the escape velocity, which in that case is only a few meters per second, a bicycle speed. We didn't think much further about it, but set the impact velocity equal to the escape velocity of the Moon, 2.4 kilometers per second, recognizing that this would require a particular category of object.

For the target of the collision, we decided to start with the present Moon, with a small iron core and a rocky mantle and crust, minus a few percent that would be represented by the impacting body out of the same rocky material but without a core. We gave it enough mass to produce a 30-kilometer-thick layer if pasted onto one hemisphere of the Moon. That equals a 1,300-kilometer diameter sphere, larger than Ceres. How about the impact angle? A head-on collision would be favorable, because that would ensure that the projectile would plunk down and splat where it impacted. But we didn't want to be favorable, so we made it 45 degrees, which is the

The Moon, as it would appear under noontime illumination from four rotating views, the farside being lower left. You can track the small Mare Moscoviense and the unusual crater Tsiolkovskiy as they rotate into view. The darkest materials are lavas that have been extruded, while the brightest materials are ejecta that radiate from recent craters.
NASA/GSFC/ASU

most probable impact angle. Would the projectile get flung around the Moon, forming a strange thick ring instead of a giant splat? We launched the run on the campus computer and went home.

THE LIMB OF the Moon is the boundary between the nearside and farside, the limit of what an observer on Earth can see. Just

beyond the limb is the crater Giordano Bruno, one of the myster-
ies of lunar geology. Only 22 kilometers in diameter, Bruno is full
of solidified magma—unexpected for a crater this size—and has
bright, fresh rays extending in all directions. Was it a small comet
impact, bringing volatiles and intense heat? Or the opposite, something large and relatively slow, so that the impact melt would stay
in the crater? Its rays are geologically young, and this led briefly to
the speculation that Bruno might even have formed in historical
times, coinciding with an event that was witnessed by five monks
on the night of June 18, 1178:

> In this year, on the Sunday before the Feast of St. John the
> Baptist, after sunset when the Moon had first become visi-
> ble a marvelous phenomenon was witnessed by some five
> or more men who were sitting facing the Moon. Now there
> was a bright new moon and as usual in that phase its horns
> were tilted toward the east and suddenly the upper horn split
> in two. From the midpoint of this division a flaming torch
> sprang up, spewing out, over a considerable distance, fire, hot
> coals, and sparks. Meanwhile the body of the Moon which
> was below writhed as it were, in anxiety, and to put it in
> the words of those who reported it to me and saw it with
> their own eyes, the Moon throbbed like a wounded snake.
> Afterwards, it returned to its proper state. This phenome-
> non was repeated a dozen times or more, the flame assuming
> various twisting shapes at random and then returning to nor-
> mal. Then after these transformations the Moon from horn
> to horn, that is along its whole length, took on a blackish
> appearance. The present writer was given this report by men
> who saw it with their own eyes, and are prepared to stake
> their honor on an oath that they have made no addition or
> falsification in the above narrative.

These statements, written down by the chronicler of Canterbury, are the data. Here are five monks swearing their testimony; it is difficult to doubt their sincerity, for the repercussions for lying are severe, in this world and the next, and there is no motive. The description that the Moon "throbbed like a wounded snake" is compelling, and moreover, the hypothesis is credible in terms of how bright a Bruno-forming event would have been. The impact kinetic energy would be equivalent to several thousand times the yield of the largest nuclear bomb ever exploded.[47] Since it impacted near the limb of the Moon, it would indeed launch a plume into space, from a viewer's perspective, that looked like "fire, hot coals, and sparks." Their reports that "it repeated a dozen times or more" would be an understandable exaggeration, as would the claim that the Moon had split in half.[48]

Still, it is incredible that there was no other record of the event—*nobody else was looking up at that time?*—so we must question the data. One problem with the monks' theory is that a Bruno-forming event in relatively recent times is highly improbable. Statistically, 20-kilometer craters form every few million years, while this observation happened a thousand years ago. There is less than a 0.1 percent chance of a Bruno-sized crater forming anywhere on the Moon in the past thousand years. And apart from the low probability, there is an even more fatal problem: that such an explosion on the Moon would litter the Earth-Moon system (*cislunar space*) with junk—something that would have significant observable consequences. The amount of escaping ejecta, over a billion tons, would find its way to Earth, making brilliant meteor showers for several hundred years,[49] lasting well into the 1700s. Those did not occur. The final inconsistency with the theory is that the floor of Bruno is peppered with small craters tens of meters across. These form at a known rate, more or less, so the age of Bruno is about a million years.[50] Bruno didn't form in historic times.

Unless the men were all hallucinating, we have to explain this remarkable and certified event. Let's expand the question to bring in more facts. We don't know a priori that they saw an impact on the Moon; that is their interpretation. What they did see was fire and coal and sparks in the sky, and they believed that it was happening to the Moon. But sometimes things do line up. Let's take a step back and consider that the Moon and the Sun appear to be in exactly the same place at the same time every once in a while, in a total solar eclipse, even though one body is actually four hundred times closer than the other. What are the odds of *that?*

IN THE EONS since its formation, the Moon has been getting smaller in apparent size as it spirals away from the Earth. As luck would have it, beginning about a million years ago, the Moon has been almost the same angular diameter as the Sun. That means that an eclipsing Moon can align perfectly in front of the Sun for a few moments, once every few years, seen from somewhere on Earth. This is a quite rare situation, a moon orbiting a planet that is exactly the same apparent diameter as its sun. A few million years from now the Moon's orbit will be so distant that it will appear smaller than the Sun, and this glorious era of total solar eclipses will end. If such a thing as alien tourists exists, then a total eclipse on Earth might be on their bucket list.[51]

Total solar eclipses take place over a narrow arc, a small (100- to 300-kilometer-wide) *umbra* (complete shadow) that races across the surface somewhat slower than the Moon orbits (1 kilometer per second) as the rotating Earth tries to keep up with the shadow. Traveling to be in the path of totality is a modern phenomenon, and I have experienced only one, thirty years ago. It changed me in a way I don't comprehend, and imprinted a new perception of who I am in space and time. I traveled with two friends, also astronomers, and a pilot whom one of them knew, and we headed south in a rented 1950s Cessna, like a Volkswagen Bug with wings, and about the same luggage capacity. We overnighted halfway down the Baja Peninsula, and on the morning of the eclipse took off early and headed to a point in time and space where the Sun, Moon, and Earth were computed to align.

As our movement through the skies of Earth progressed, I felt like a mouse climbing through a giant clockwork, gears meshing and pendulums ticking, the gong about to go off. A pencil-beam shadow cast by the Sun and Moon was racing toward the Earth through space, and we raced to meet it in our flying contraption, piloting the convecting skies on a beautiful clear day, the mountains to our right, the ocean below. Others had the same idea, and my job was to monitor for other eclipse-bound planes—we could see six at one point. An hour later we touched down on a dirt airstrip behind a small beach hotel, taxied to where twenty or thirty other planes were parked, and walked through a charming, simple courtyard and down the steps to a beautiful beachfront, where dozens of other time and space travelers had already converged, waiting on the foretold miracle.

As I was to experience the next hour, you don't *see* a total solar eclipse. That is the folly of taking pictures of it—let somebody else do it, who is much better at photography and has tens of thousands of dollars of equipment. I'm so glad that there are photographers, especially, now with high-definition movies that convey the stunning movement of the corona.[52] But no matter how articulate the image, it will never convey a fraction of what you experience by just gazing up with your two eyes and all your bones.

In Arthur C. Clarke's novel *2001: A Space Odyssey,* there is a scene where Moon-Watcher, the leader of a tribe of early hominids, tries to hold the Moon between his finger and his thumb. There is a sound of crickets. The next morning the clan wakes up to a tall, black, gleaming rectangle in the middle of their camp;[53] they jump and scream and touch it; it blows their minds. A total solar eclipse is like that, but with less screaming and exulting these days. Our enlightened minds know exactly what's happening. We've planned long in advance to be experiencing exactly this. A prehistoric human clan would have no forewarning or any previous experience.

Once the Moon starts to move in front of the Sun, for the next half hour the impression is that there is some high-stratus cloud cover that grows darker. Even a sliver of the Sun is too brilliant to look at, so you don't see the tiny bite that is being taken out of it at first. Your eyes adjust. But after half an hour you perceive that everything appears curiously *sharper*. That's because the Sun is not a point source of illumination, it is a distributed source, the size of your little fingernail at arm's length. Indeed, sunlight going through a pinhole doesn't cast a pinhole shadow, but projects an inverse image of the Sun. (You can try it.) As a result, the edges of sunlit shadows are always a bit fuzzy, in pinhole photos and to astronauts walking on the airless Moon.

As the Sun turns into a thin crescent shape, the shadows sharpen, but only in the skinny direction. In practice it doesn't affect anything you do, because there is still plenty of light, but your visual cortex, designed to notice contrasts, wakes up to the novelty. Things get a little surreal. You go to rest under a tree and notice thousands of little crescent-shaped shadows of the Sun, each one a little pinhole photo. If you squint and peek, you see that the Sun has been reduced to a brilliant white scimitar. The Sun is being eaten by the Moon!

Pelicans circle beyond the shore, spiraling up into a sort of question mark. An old man is standing on the rocks where gentle waves are crashing. We're lying in the sand with our mirrored glasses and, for when totality begins, our binocs.[54] There's a restless energy about, as though someone has spiked the punch with something

From the total solar eclipse of August 21, 2017, seen from outside Jackson, Wyoming. This image used exposure bracketing to show both the Sun's co-rona and the surface features of the new Moon itself.
Converted to gray scale; original photograph by Michael S. Adler (CC BY-SA 4.0)

strange. And then it happens. "Look at the sea!" calls out the old man. We stare out beyond him, and the open sea stretches out like an elastic infinity, still cloudless but suddenly alive with color and transformation. The last sliver of sunlight refracts over the tops of the Moon's mountains, making patterns that race across the Earth as waves and purple bands. And then these optical phenomena all race to catch up with one another and . . .

I can no longer describe what I saw, but I remember the *sound* that it made—a giant gong. Then everything was still. The waves kept lapping, the birds had dispersed. The planets came out in the strange new twilight, and the brightest stars. And there, blazing above our heads for six beautiful minutes, was the black Sun with its fiery crown, like someone had plucked a hole out of the sky.

ECLIPSES ARE MAJOR astronomical events throughout the world, and the earliest records are intermingled. In the fourth century BC, the time of Aristotle, Gan De wrote several lost works about planets. He may have spotted Ganymede as a bright star next to Jupiter. Fast-forward to a time when Western science was at its nadir, a century before the monks of Canterbury, and the polymathic statesman Shen Kuo composed *Brush Talks from Dream Brook* while retired to his tranquil gardens.[55] Completed in 1088, it is an edifice on music, geology, astronomy, metallurgy, botany, medicine, even UFOs. Seven centuries before geology would become a science in the West, Shen Kuo understood the fossil record and its implications for deep time.[56] Strata of marine fossils in a high cliff implied the rise of the land or the retreat of the sea, he wrote, and petrified bamboo in a desert outcrop was evidence of climate change. He also knew the Moon is a sphere, illuminated by the Sun: "If half of a sphere is covered with powder and looked at from the side, the covered part will look like a crescent." On why eclipses do not happen every full or new moon, he explains: "The ecliptic and the moon's path are like two rings, lying one over the

Sun, Moon, branches, and Earth.
Meryl Natchez / http://www.dactyls-and-drakes.com

other, but distant by a small amount."[57] We have to wait for the cycles to align.

Fireworks going off on the Moon later in the Song dynasty would have been faithfully recorded; this was the time of the invention of gunpowder and paper currency. So what *did* the baffled monks see, the day before the Feast of John the Baptist? Well, that happens to also be the season of the Taurid meteor shower, an event that we now know is caused by the Earth passing through the orbital debris of the periodic comet 2P/Encke, which happens regularly every year. So there's an alternative hypothesis to the monks' story: one thing in front of another, instead of on another: a bright shooting star that headed straight at them,[58] right in front of the Moon from their viewpoint.

It sounds unlikely, but not really. I've seen one meteor in my life coming straight at me, and it appeared to meander almost as if it were drunk, as I was right along its vector of approach. There have been millions of backyard gatherings worldwide in the past thousand years, so it is actually likely that this would have happened to someone, a bright meteor coming directly at them in a geometry that's right in front of the Moon.

We exist in space and time, and perception defines what we see. Sometimes we jump to conclusions; other things are obviously real. An alien living on a planet orbiting a star high above the Milky Way, looking down on the huge galactic spiral every night, would *know* from the start that there is a great big Universe beyond us. Here in the crowded midplane we have required centuries of astronomy to figure it out. Or consider how the universe might be originally perceived from the surface of a planet in a tightly packed system like TRAPPIST-1, where adjacent planets swing close enough that they loom in one another's skies like full moons, except they don't orbit each other. They make looping ellipses around the sun, and close approaches as they race ahead and fall behind (as you rotate on your own world, whose little moon goes around you . . .). As the days and weeks advance, equaling planetary years, this spinning teacup ride might yield an incredibly rich sky mythology but frustrate a physical understanding of astronomy.

The Earth-Moon system was once its own wild ride. If you could stay in one spot on Earth and rewind the clock in a time machine, you'd observe the Moon coming closer and closer, growing larger. It would go through its phases—quarter moon, full moon, new moon—orbiting faster. Days would grow shorter as Earth got its angular momentum back. The sloshing tides would grow higher until at some point they would overlap the coasts; the earliest landmasses would have had a rough time.[59] Further back, as far as you can go, and the Moon would be where Darwin first imagined it, beyond the tipping point of 3 Earth radii where an orbiting body spirals out instead of in.

To go further back in time is the giant impact itself, so let's stop here awhile and enjoy the view. The Moon, at this distance, would be orbiting every six hours, a bit slower than the Earth was spinning then, which is why it spirals out. From where you are on Earth, the Moon would barely crawl across the sky—almost *geostationary*—although it would race against the background stars each night.

Solar eclipses by the Moon would happen regularly. A month, measured from full moon to full moon, would happen every six hours at first, increasing as the lunar orbit expanded; the Earth day would lengthen, its rotation slowed by tidal coupling. During this close early dance, Earth-Moon resonances with each other and the Sun would have influenced the plane of the lunar orbit, and might have sapped the Earth-Moon system of its angular momentum.[60]

If you looked up from the Earth at the onset of the Hadean, the Moon would be the size of your palm, so big that you would behold it in geological detail the way an astronaut might look down on the surface of a planet. Dark fractured masses would glow red with eruptions and volcanoes, punctured by collisions, and blocks would pile up into bands of gray mountains. The Moon would be circled by a band of rings extending a few diameters, inclined by a few degrees, like Saturn's but composed of rocky minerals instead of ice. Subject to the gravitational whims of Earth, these rings would go away, but be replenished by impacts until the final sweepup was over.[61]

Before there was even a Moon, there was a disk of about two lunar masses around the Earth, the remnants of the giant impact. This far back, there would be no place to sit down; the crust would all be molten, sweltering beneath a silicate-vapor atmosphere. But out in that disk, things were all about to change: it became gravitationally unstable. Most likely there was a substantial lump of iron in the disk, a fragment of the core of Theia in the standard model, or a Vesta-sized object that accreted out of the densest metallic blobs. This giant lump of iron, three times the density of rock, would rapidly capture other material from the disk, soon becoming 1,000, then 2,000, then 3,000 kilometers in diameter. Its presence would substantially alter the gravitational dynamics around the Earth. In addition to sweeping up material, it altered the *potential field*, creating the possibility for Trojan moons around the Earth.

You may be familiar with the analogy of gravity being represented as a dimple in space-time created by a mass on a rubber sheet. The Sun is like a heavy cannonball in the center of the sheet, and the planets orbit around this curvature of space-time, like a nickel rolling down the coin-funnel game at the airport. Gravity is the gradient of the potential field. Jupiter, then, is like a steel marble that makes its own dimple as it goes around the Sun. Now you have two dimples orbiting each other.

Let's say you want to plot these dimples, and for simplicity you always plot them with Jupiter to the right of the graph. (That's okay, as long as you remember the graph is rotating.) You can write down the laws of physics in this rotating frame by factoring in the merry-go-round physics known as *Coriolis forces*. A ball (i.e., a planet) that you throw straight ahead on such a plot (or merry-go-round) doesn't go straight—it curves. When you add the gravity forces—which are there all along—plus Coriolis forces, then in addition to the two big dimples in the rubber sheet, there are two smaller dimples 60 degrees ahead of and behind the planet known as the *Trojan points*, L_4 and L_5. There are three other stability points in this rotating frame of reference, but instead of being dimples, these are plateaus and saddle points, L_1, L_2, and L_3 (collectively, the Lagrange points).

The Trojan points can trap material, although the trapping can be temporary. Jupiter has over seven thousand Trojan asteroids discovered so far. Mars has a handful, most of them trailing at L_5. Neptune also has a handful of Trojans, mostly leading at L_4. Earth has only one known Trojan asteroid, a few hundred meters in diameter. Venus has one; it's on a wide-looping "tadpole" orbit, probably unstable. The lack of populations of Trojans at Venus and Earth is telling . . . of something.

Trojan points can provide small objects with protection. Asteroids and comets that end up orbiting the Sun at around 5 AU are likely to crash into Jupiter or be ejected unless they get captured as Trojans, to live in a Sargasso Sea bounded by strong dynamical currents. Jupiter's Trojan population is the focus of NASA's upcoming Lucy mission, where the primary science question is whether they are original pieces from Jupiter's formation or are captured riffraff from much later—in either case a trove of information about how our planets came to be.[62]

This brings us back to Moon formation. Think of the Earth and this massive proto-Moon that has accreted about 3 Earth radii away. Plot these on the same rotating graph, now putting Earth at the center and the proto-Moon to the right (where you plotted Jupiter before). The proto-Moon is very close to Earth, and 1 percent of its mass, so its Trojan points L_4 and L_5, the two extra dimples on the plot, will be substantial. And so begins the story of two moons.

The Earth has finished forming, but is a roiling mess, and the Moon is rapidly accreting from the protolunar disk. Small bodies are getting trapped around the Earth-Moon Trojan points. Saturn, being the only planet to currently have Trojan moons, gives the only example of what to expect. Calypso and Telesto differ by a factor of 2 in mass, and share an orbit with Tethys. Helene and Polydeuces differ by a factor of 2,000 in mass and share an orbit with Dione. So for the Trojan moons of the Moon, maybe they differed by a factor of 10 to 100 in mass. Importantly, they would be accreted out of the same feedstock as the Moon, but without any iron.

This scenario of two moons is based on reasonable scenarios for the outcome of giant impacts, but you still have to choose some parameters. The Trojan moon would have to be about a third the diameter of the Moon (1/30 the mass) to account for the added farside crust, and a second Trojan would likely be several times smaller than that,[63] subject to disruption by the hailstorm of giant impact remnants that would sweep through the Earth-Moon system over the next million years. It would be far less geologically significant. Tens of millions of years later, the largest had survived, and had solidified. They migrated hand in hand farther and farther from Earth—until one day they drifted too far and fell out of their reverie.

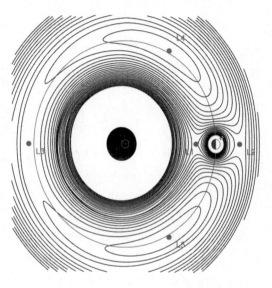

MARTIN OBTAINED THE computer modeling result a few days later, and showed me a low-resolution version of what you see here, minus some tweaks. We could not guess that our result would soon go viral in our hometown newspapers

A moon that shares an orbit around a planet influences its gravitational potential. When plotting with the moon always to the right, then you have to add in the Coriolis (merry-go-round) forces. When you do that, the lines show contours of constant potential; they look like gravity field contours except there are dimples at the Trojan points (L_4 and L_5, above and below) where matter co-orbiting with the moon can be trapped. There are other Lagrangian points—for example, L_1 in between the two bodies, and L_2 just beyond the Moon.

Image generated by the author using gnuplot

and international news magazines; for now we were just having fun figuring something out. We stared at the result. It reminded us of our models of comet splats, in that friction appeared to hold the projectile together. But for the larger mass, and the greater speed of this giant lunar collision, the projectile flattened itself out into more of a pancake. Most important, *it stayed in the hole!* We had a billion-car pileup on our hands.

Our idea was now a hypothesis, so we had to think more carefully how to validate it, how to challenge it. The worst thing is to try to publish something that ends up being obviously wrong, so you first try to prove that it's wrong. Our idea worked well only because we had chosen a low-velocity collision. A faster collision would have penetrated deeper, producing an impact basin like SPA and not a splat. So that ruled out comets and asteroids. The only possible projectile would be another moon of Earth, as anything from outside the Earth-Moon system would be zooming by at 10 to 20 kilometers per second.[64] Would that be feasible? We liked the idea of a Trojan moon because its stable orbit would fall apart after around ten million years, as the Moon migrated away from the Earth, like the fuse on a ticking time bomb.

If you are able to put your proposed events in time, then you have won half the battle of developing a hypothesis. To explain the geology, the proposed collision of the Trojan moon with the Moon would have had to occur after they both were mostly solid. According to cooling models, that takes about one million to ten million years. That worked out great, in terms of timing, because after around ten million years (according to tidal models) the lunar orbit would have expanded to about 20 Earth radii, at which point the "Laplace plane" of the dynamics becomes governed by the Sun. The "dimples" in the rubber sheet are weakened, and the Trojan moon drifts free.

Later that day Martin plotted up the modeled topography compared to the lunar topography; it matched the profile of the fos-

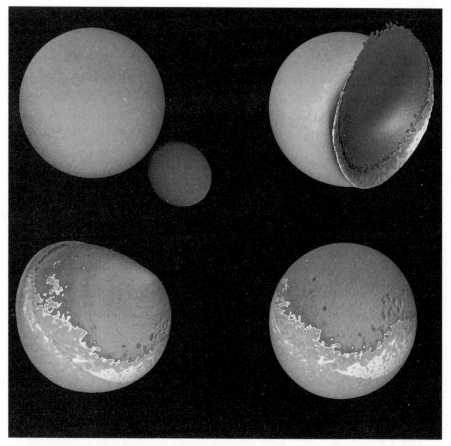

Four snapshots of the impact event that formed the lunar farside highlands, according to Jutzi and Asphaug (2011), shown at initial contact and at 0.6, 1.4, and 2.8 hours. The 1,200-kilometer diameter projectile makes a crater, sure enough, but then it overflows its crater. The lightest material represents a liquid magma ocean underneath the solid crust that gets squeezed over to the other side, contributing the KREEP anomaly.

Martin Jutzi, University of Bern, Switzerland

silized highlands bulge just as well as the frozen-in-tidal model, without having to explain the missing nearside. *Wow! It worked!* We hadn't put a lot of thought into the astrophysical or geophysical or celestial dynamical implications, but we would now have to.[65]

This was a new kind of calculation—giant solid planets with

granular friction, colliding slowly—so we needed to check our result. We had over the years validated the basic physics in the code
of porosity, crushing, friction, and fragmentation, so we believed
our models—to a certain extent—but were pushing the envelope
way outside the laboratory. We needed something to compare to, a
benchmark. There were the tried-and-true crater scaling laws that
applied to flat, semi-infinite planets; if you plug in the numbers for
the Moon, a 1,300-kilometer projectile hitting at 2.4 kilometers
per second[66] displaces only a fifth of the volume of the projectile.
So the simulations agreed with that prediction: a projectile that
overfilled its own crater.

Encouraged by this basic validation, we decided to be more re-
alistic, increasing the resolution by a factor of 3, committing to a
much longer computer run time. As an afterthought, just before
launching the run, Martin had the idea of adding a 10-kilometer-
thick melt layer beneath 30 kilometers of solid crust, to represent
the deep, residual magma ocean, the KREEP. Physically it wasn't
much different—the same substance, only hotter—but it gave the
story an important twist. KREEP is prevalent on the nearside and
almost absent on the farside. The concentrated radioactive elements in KREEP, including uranium and thorium, are thought to
have driven the relatively late heating of the magmas that flooded
the lunar maria. The smaller moon, we made out of a single material with the same composition as the bulk lunar mantle, only
slightly less dense as it was less compressed by its gravity.

We had some decisions to make: choosing the projectile and
target properties and the collisional setup parameters. Not only
do supercomputers cost time spent waiting for answers (weeks to
months) and money (a run can cost tens of thousands of dollars),
but once you have spent all that money and time, you can become
locked into your hypothesis—in for a penny, in for a pound. In
addition, once you start making the beautiful graphics, you can

become enamored of your creation and lose your objectivity. Our higher-resolution simulation took ten days to complete. We did not expect that the result would change very much compared to the pilot study, so while it chugged away, we started putting together the manuscript and thinking up the dynamical scenario of Trojan moons.

The new simulation that tracked the KREEP layer gave us an unexpected surprise: the splat squished the KREEP, like mashing your fist into a cherry pie, and the filling got pushed to one hemisphere. Because it contains a radioactive heat source, this concentrated KREEP could explain why massive volcanic flooding is limited to the lunar nearside, with its regions of high heat flow and young maria. The farside, on the other hand, became geologically dead, like when you drop a shovel full of cold dirt onto the coals of a campfire.

The hypothesis was invented to explain the twice-thick farside crust, but would also explain the absence of farside KREEP and the dichotomy in lunar geologic activity. Unfortunately, the hypothesis has been able to slip away from the tests that would directly challenge it. Gravity alone would not detect it.

The farside highlands are, as far as we can tell, made of basically the same stuff as the nearside highlands. Although we have never sampled the farside directly,[67] of the hundreds of lunar meteorites, about half of them (statistically) come from the farside, but you can't tell which came from where. Because the Trojan moon formed out of the same protolunar disk as the Moon, composition isn't much of a test—it would solidify from a magma ocean made out of the same stuff, although under a fraction of the internal pressure. Closely similar rock sequences would result, with a denser olivine-rich interior and a plagioclase-rich exterior, a mini-Moon but without the core.

Although we didn't include it in our simulations, the 1,300-

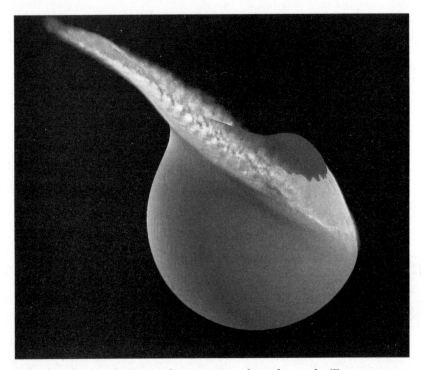

A few hundred seconds into the two moons hypothesis, the Trojan moon is becoming a hemispheric splat on the target moon.
Martin Jutzi, University of Bern, Switzerland

kilometer diameter moon would have an olivine-rich interior that would end up sandwiched in the middle of the pancake, ending up as an olivine layer about 10 kilometers below the farside surface after the splat was finished. (No iron, though—that would have been swept up in the Moon's core, the initial act of accretion.) There are a lot of weird and unexplained exposures of olivine on the Moon, most clearly mapped by Japan's Kaguya orbiter, which would be consistent with a zone of olivine that's relatively shallow, accessible to Chicxulub-sized impacts. Our model also predicts we would find some olivine exposed along the nearside-farside boundary (aka the limb), although after 4.4 billion years, such exposures would be indistinct.

Another test is geophysical. A collision at 2 kilometers per second is too slow to produce a strong shock wave, according to laboratory studies of rocky materials.[68] But it's violent enough to do a lot of damage. The model predicts that there should be a planar *shear zone* where the original surface of the Trojan moon came to a screeching halt against the surface of the original Moon. This contact zone would be crushed and deformed intensely, creating tens or even hundreds of meters of friction-melted rock buried 30 to 40 kilometers deep. That boundary, if present, would show up clearly in seismic reflection images, so I hope there will soon be opportunities to disprove it!

CHAPTER 7

A BILLION EARTHS

THE BRANCHING STRUCTURE OF PLANETARY accretion is a tree of growth. The smallest planetesimals are leaves, and larger planetesimals, made out of those, are stems. Theia-sized embryos are branches, leading to the trunk, which is the planet. Instead of looking at the specific detailed pathways that might have brought us to this place, let's think of the bigger picture, and for that let's consider further analogies.

However they are accreted, big planetary bodies are harder to catastrophically destroy than small ones; that's intuitively clear. They have a lot more gravity and a lot more mass to blast apart. Models of collisional evolution show that asteroids larger than about 200 kilometers in diameter are probably relatively intact nuggets of planet formation—they are so massive that a collision energetic enough to disrupt them would be unlikely. Major asteroids, like Psyche to be visited in 2026 by the eponymous NASA mission, are frozen in time, much like the Moon has been, subject to a hail of projectiles over billions of years, but none big enough to completely destroy them.

Asteroids smaller than 100 kilometers in diameter, on the other hand, are *likely* to have been catastrophically disrupted by collisions according to calculations. There were so many more of them early on that their collisions with each other were frequent, and being smaller they were more easily disrupted. So it's a tipping point that depends on where you are in the process of planet formation, whether the objects of a certain size tend to get bigger, winning out, or tend to get smaller, eroding and being blasted. Today the Main Belt is slowly eroding away, collision by collision, but 4.56 billion years ago in that region, small bodies were accreting and growing.

As for asteroids smaller than 100 kilometers in diameter, these are thought to be the result of *collisional grinding,* the production of innumerable 10-meter bodies, billions of 100-meter bodies, and millions of kilometer-sized bodies, when rocks break apart. The analogous experiment is to start with large rocks, dump them into a cylindrical grinder, and turn it on. A lot of dust is generated at first, and the weak rocks are the first to tumble apart. Eventually everything will become dust, but in the meantime most of the mass is in a few large fragments. Today the planets have cleared out—the power of the grinder is much weaker—and the mass is mostly in the big nuggets, with smaller bits slowly breaking themselves down.

Half of the mass of the present asteroids is in four bodies, sometimes called *dwarf planets*: Vesta, Ceres, Pallas, and Hygiea, ranging from 400 to almost 1,000 kilometers in diameter. That's not too surprising, seeing as half the mass of the terrestrial planets is in one body, the Earth. As we have seen, accretion creates a top-heavy mass distribution. Vesta, Ceres, and the other big ones are, I think, originally accreted bodies, or immediate remnants thereof. Then there are dozens of asteroids that are a few hundred kilometers in diameter, a few of which might be primordial, and hundreds that are half as big as that, and so on. The size distribution becomes a geometrical progression, where for every asteroid there are several smaller ones that are about one-tenth as massive,[1] in the same way that a clay shooting target breaks into several recognizable pieces that you could put together (asteroid families, that is), and ten shards, and hundreds of bits, and thousands of particles, and ultimately dust.

This hierarchy of fragmentation can correspond to a hierarchy of surface and near-surface zones, with volatile exteriors interacting with all these self-gravitating broken bits. Topmost is the optical surface—the outer micron that reflects and refracts sunlight into a camera lens. It's what a picture shows you. Beneath that first micron, the camera itself knows nothing. Next is the thermal surface, the zone that feels the presence of the Sun's heat, extending about a centimeter down on daily time scales, and meters down on annual time scales. We bury our root cellars and wine cellars below the annual thermal layer, and you wiggle your toes beneath the scalding sand to cool off, beneath the daily thermal layer.

Below the optical and thermal boundary there is the near-subsurface, the zone that communicates with the outer atmosphere, or if there is none, the radiation of space. On planets with atmospheres, the near-subsurface is where water (vapor and liquid) is absorbed by the soil and exchanged with the air. On Earth, it's where most of the biomass is found. On an airless body like a comet, the near-subsurface extends to where easily vaporized ices[2] outgas into jets and tails. On Mars, it extends meters into an upper regolith that inhales and exhales seasonally, exchanging H_2O and CO_2. On Neptune's moon Triton, it extends to the source of the nitrogen-driven geysers seen during the Voyager flyby.

Looking at primordial nuggets (e.g., Psyche, Vesta, Ceres) of incipient planet formation, plus the disrupted fragments (smaller asteroids and comets), we hope to reassemble some of the first clay pigeons. But now imagine that you have been given a random box with only a fraction of a percentage of the fragments. What story would you put together? What kind of pigeon? There were probably thousands of times more asteroids of all sizes to begin with, so most of the pieces belong to long-gone puzzles.

For asteroids smaller than 100 kilometers, disruption *is* creation. Each is derived from the breakup of a larger parent body, so their formation is hierarchical—breakups at the top causing a cascade of fragments. For bodies larger than 1,000 kilometers, on the other hand, accretion is creation. Collisions cause mergers, growing embryos into oligarchs into planets. This is also hierarchical, but fed from the bottom, a tree of growth. How this actually plays out is not understood—hence the importance of upcoming NASA missions to intermediate-sized asteroids like Psyche and Patroclus[3] that are at the cusp.

Because it accreted hierarchically, we can't specify the time of formation of the Earth. You can determine how much time went by from t_o until the Earth's core was separated from its mantle,[4] but this is a lower limit on how long it took the Earth to form, since iron might have separated out inside of the smaller embryos from which it formed. And then the Moon-forming event appears to have happened about fifty million years after *that*, consistent with it being one of the last accretions of the oligarchs.

Geologically speaking, after Theia collided, Earth was a brand-new planet, recycled from its old components, although there is a lot of debate as to how well mixed it got and whether big lumps of Theia are left "down there" in the mantle, compositionally intact. That gets into the energetics of the collision and the specific Moon formation scenario: a hugely energetic giant impact melts everything and hits the geologic reset button. A gentle merger might leave remnants of Theia as layers inside the deep Earth.

Once the Earth's crust solidified following the Moon-forming giant impact, geology began, transforming the planet too rapidly to keep track of at first, and then thickening into an encrusted sluggish mantle, like a porridge on the stove. Large bodies continued to collide and sometimes destroyed this early crust, so there were many false starts. Unlike the massive Earth, which remained partly molten, the Moon solidified within 10 million years, according to thermal models.[5] Ironically, the more the Moon got bombarded when it had a thin crust, the *faster* it cooled; each impact would peel off some crust, exposing the magma ocean, stirring up the porridge, allowing heat to escape. In that case, the Moon could have become solid in only a million years,[6] other than the KREEP layer with its radioactive heat source.

Geologists don't work with linear time unless they have to—so long as they can place events *in* time. Thus, since the pioneering work of Gene Shoemaker, we have been striving to correlate geologic time, as a number anchored by events, across the solar system, looking for smoking guns and holy grails. The obvious place to begin making these connections is in our own backyard, on the "seventh continent," the Moon. Because the Moon was so quick to lose its energy for transformative geology, it is there that we might find the best-preserved remnants of the tribulations that Earth was going through soon after its formation.

The Nectarian and early Imbrian eras on the Moon, 3.5 billion to 4 billion years ago, were coincident with the first profusion of life. Gigantic bombardments also happened on the Earth, and although the resulting impact basins were swallowed and transformed by the overturning geology, they ejected huge amounts of crustal material into space. Just as the Earth gets hit by meteorites ejected from the Moon, the Moon was bombarded by rocks ejected from the Earth, where they got mixed into the lunar regolith—small rocks and grains and giant boulders. They would have landed in profusion around the time of the origin of life, most of them at a relatively gentle 2 to 3 kilometers per second, a bit above the lunar escape velocity. There's little hope that Earth organisms would have colonized the Moon, but signatures of original life might be preserved.

Zircons, you may recall, are high-temperature silicate minerals whose laboratory analysis reveals radiometric clocks and pressure-temperature records of their formation. On Earth, zircons inform us about the Hadean, especially, the comings and goings of early planetary volcanism and what conditions were like—including oxidation state, the amount of free O_2, which is related to the presence or absence of water.[7] Zircons are also found in lunar samples, and most of them are native to the Moon, based on their chemistry. But Apollo 14 sample 14321 may be from Earth. It has quartz and feldspar clasts that contain zircons that appear to have crystallized four billion years ago under oxidizing conditions,[8] temperatures and pressures quite different from other lunar zircons, more typical of the fluid-rich low-temperature environment that was common beneath the early continental crust of Earth.

So that's the tip of the iceberg. Somewhere beneath the astronauts' boots were larger nuggets of more interesting, better-preserved Hadean Earth materials waiting to be found, a sample return mission in space and time. But the search for Earth rocks on the Moon would be intrinsically aimless, the task of a robot that sieves and sorts for years and is incapable of boredom. Meteorites can be anywhere, like photos in your grandma's attic,[9] dusty relics from a time you never knew about. But you have to start opening the boxes.

WE LEFT OFF, a while back, on the topic of planetary diversity. It makes me think of Leo Tolstoy's famous line, the opening of his novel *Anna Karenina*: "All happy families are alike; each unhappy family is unhappy in its own way." Statisticians have expanded on this a bit, referred to as the Anna Karenina principle: if deficiency in any of several factors causes failure, then success requires there to be no deficiency in any of those factors. Applied to the topic at hand, we might say: *All accreted planets are alike; every unaccreted planet is unique in how it was not accreted.* What is an unaccreted planet? The poster child is the planet Mercury; the poster pups are the incredibly diverse asteroids and other relics of the Main Belt—they're anything that's left, once the winners—the accreted planets—have taken almost all.

According to the Anna Karenina principle, a planetary embryo can have any number and any kind of encounter with a larger planet, provided that in every encounter it avoids being accreted. Either that, or it never encounters a larger planet that would accrete it—either by itself being the largest planet around, or by having a dynamically isolated orbit. The Sun accreted more than 99.8 percent of the mass of the solar system, and Jupiter accreted more than 70 percent of the remainder, so every planet is lucky in that sense. It takes further luck for any object in the terrestrial zone not to have been accreted by the Earth or Venus, which accreted over 93 percent of the rest. Mars, Mercury, the Moon, the asteroids—they add up to only 7 percent of the total, so when you look at these objects, you have to marvel at how unlikely they are to exist.

In the Main Belt, the most massive asteroids like Vesta, Ceres, and Psyche are as diverse as bodies can be—rock worlds, ice worlds, metal worlds. If they formed in the same region of the solar system and grew by feeding on planetesimals to become the largest bodies in the Main Belt, wouldn't you expect they would be at least *somewhat* the same? If instead each of them was subject to the whims of luck, like the few surviving soldiers in the analogy above, they would be diverse, but it would imply that there used to be a major planet roaming in the Main Belt that accreted *almost* all of them—all of them but these ragged survivors. The planet? It disappeared.

Mercury may be the clearest example of a surviving soldier, who avoided being accreted by the Earth and Venus. Spacecraft measuring Mercury's gravity have determined that its iron core is four-fifths of its radius; the silicate crust and mantle are like the frosting on a cupcake. The cores of the rest of the terrestrial planets (Earth, Mars, and Venus—the Moon being the other weird exception) are only half the radius, about 30 percent of the mass. How did Mercury lose almost all of its rocky mantle? You could destroy proto-Mercury, as proposed by Swiss astrophysicist Willy Benz,[10] one of the original researchers into the giant impact theory, but this led to the problem that Mercury would sweep up almost all its mantle, re-accreting it quickly as it orbits around the Sun. Plus, it would eliminate all its volatiles.

Once as a PhD student I was learning to run Benz's simulation code, and just for fun I set the parameters for two differentiated Vesta-sized asteroids colliding at twice their escape velocity, a few hundred meters per second. It was a hit-and-run collision. I remember thinking how beautiful that was—two planets, first whole, then leaving with their mantles strung out into spirals. Later, when I worked with astrophysicist Robin Canup on the problem of Moon formation, we were simulating Theias colliding with Earths, and I thought it was so cool how many of these Theias kept going, worse for wear but recognizable. These runs didn't give us a massive protolunar disk, so we pressed on.[11]

I kept thinking about all those lost Theias. Science follows missions and big instruments, and in 2011, the planet Mercury was the biggest mystery presented at all the conferences. The MESSENGER mission was obtaining fantastic data sets, transforming a sketchy black-and-white little planet into an incredibly rich palette of color images, mapping various data products into red, green, and blue at high resolution. (Visibly, the planet is gray to human eyes.) The biggest mystery was the clear evidence for an abundance of volatiles on and inside Mercury: complex "hollows" in the crust where huge scarps have back-eroded, and ground ice in the permanently shadowed regions, and the detection in its rocks of a much greater abundance of the semi-volatile element potassium than the Moon has. A hot, airless world, thought to be born in some kind of giant impact, Mercury is not supposed to have volatiles.

What if Mercury was the result of a giant impact, not *by* something, but *into* something, like one of those Theias that kept going? The Moon is a giant impact by-product and ended up with far fewer volatiles, but a hit-and-run collision is different, using the gravity of the bigger planet to help remove the mantle of the smaller, so I was intrigued by the possibility. You don't just gracefully remove half the mantle from a planet. Halfway inside of proto-Mercury, prior to its collision with something bigger, the rocks would have been under extreme pressure; after a hit-and-run collision, the interior would decompress and all kinds of igneous geochemistry and degassing and abrupt cooling could take place.

There were two problems with the theory. One, the seeming low probability of that. If you collide proto-Mercury into proto-Earth/ Venus (let's say it's Venus), then it is likely to impact Venus again. If it eventually accreted with Venus, which would seem likely, then Mercury would be gone and we wouldn't be talking about it. But if it collided and kept going—a hit-and-run collision—then it would be one of those next-largest sharks that got away, and got away again—however unlikely that may seem, here it is.

But it's not really that unlikely; it's just unlikely enough to explain

Mercury's weird geology. Based on our simulations, a hit-and-run collision happens around half the time, whenever a proto-Mercury-sized body hits a Venus-sized body. So Mercury had to be lucky once, twice, maybe even three times, until it got back on track and stopped colliding with planets. (Things eventually did settle down.) That's like flipping a coin heads three times in a row, a one-in-eight chance ($1/2 \times 1/2 \times 1/2$). Since proto-Earth and proto-Venus were built out of eight or more other, less fortunate proto-Mercurys, only one of them had to get away, so it's actually *probable* to have one planet like Mercury.

Mars could have survived a different way. Instead of surviving every giant impact it had with a larger planet by having a hit-and-run collision, Mars may have had close calls but never collided with a larger planet, colliding only with bodies that were smaller than itself. Maybe it was dynamically isolated; maybe it was lucky. While it's tempting to embellish the story with further details—this is why Venus is so different from the Earth and so on—chaos and caution advise us not to go there until we have samples from Venus. But the model makes predictions that can eventually be tested: the largest planets in a system should grow to have about the same bulk composition, while next-largest planets should end up with enhanced compositional diversity. If you ground up Venus and ground up Earth and sampled a beaker of each, you'd be hard-pressed to tell which was which. Not so Mercury, Mars, and the Moon; or Vesta, Ceres, and Psyche.

WE COMMONLY HEAR that the Sun is an average star. But that's not quite true; the most common stars, called *red dwarfs,* are much smaller, big enough (tens of Jupiter masses) that they ignite hydrogen fusion in their core, but small enough that they don't put out as much light and heat. This makes them hard to discover. Indeed, the closest star to us is probably not Proxima Centauri, but some undiscovered red dwarf.

If you are a planet wanting to incubate life around such a star, you have to be close to the fire, orbiting only a few hundredths of an AU away. This makes them a great place to hunt for habitable planets: there are many such stars, and their planets would orbit so close that under the right geometrical conditions, you might see them obscure some of the starlight now and then. All you have to do is stare and stare and stare—and stare at a lot of these stars, because we have to be lucky enough to be in the *ecliptic* of their solar system to get a star shadow, a subtle dimming of the starlight while the planet passes in front, what we call a *transit*.

A few years ago, a consortium of astronomers in Belgium and Chile pointed a modest twenty-four-inch telescope to stare at several nearby "ultra-cool M-dwarfs" in the hope of finding a system of planets. It was a very successful hunt. The first of their finds, a solar system they call[12] TRAPPIST-1, is a treasure trove with seven planets. From the amount of light each planet obscures when it transits, we know that they are each comparable to the Earth in size. Five of those Earth-mass planets orbit within the estimated habitable zone of their star, where water can be a liquid on the surface, putting the system on the top of the list for the search for alien life.

The discovery data for TRAPPIST-1 are just a bunch of dots, so let me walk you through them. Although we can't directly observe these planets, even with the most powerful telescopes, we are "edge on" to their solar system, so each planet winks in front of its sun every one of its years, which is only 12 Earth days for the outermost and 1.5 Earth days for the innermost, because the planets orbit so close to their little star. When they pass in front, the starlight dims by a few percent for a few hours. As the week-long years go by, astronomers get more and more data, improving

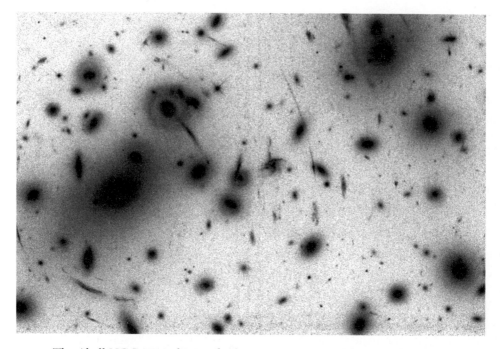

The Abell NGC 2218 cluster of galaxies is so massive that its gravity bends the light from more distant galaxies, creating cosmic lens flares. (To render the color image into gray scale I took the old-fashioned approach of making a negative image, which has a way of bringing out certain details.) *NASA/HST/A. Fruchter*

the models by discriminating *signal* (that which is real) from noise inherent to any one observation.[13]

Transits are the most powerful observational tool for characterizing distant planets, but we see a transit only if we are in the orbital plane of *their* solar system. Otherwise the shadow misses us, in the same way that the Moon is constantly casting a shadow of the Sun, but only occasionally does it land on Earth. The converse is also true: alien astronomers need to be in the Earth's orbital plane (our *ecliptic*) if they are to learn anything detailed about the Earth. The vast majority of alien astronomers, those living on planets around stars that are north and south of *our* ecliptic, would be unlikely to know about the existence of Earth. But to astronomers along our ecliptic plane, especially those within a few hundred light-years, the Earth's shaded cutout in front of the Sun, beaming through space, might give them enough information to discover that one-fifth of our atmosphere is oxygen; from that they might infer that we have life.

The mass of the TRAPPIST-1 star is derived from Kepler's law applied to the orbiting planets; it is eighty-four times the mass of Jupiter, one-eighth of a solar mass. From the duration of the transits, we know that the star is 50 percent bigger than Jupiter—compressed to an average density ten times that of iron, so dense because its internal fires are relatively weak, so it doesn't inflate thermally the way large active stars do. The sizes of the TRAPPIST-1 planets are also known from the transits; and in turn, their masses can be estimated by detailed orbit-fitting, since the star and the planet tug on

Composite of images of the Venus transit taken at extreme ultraviolet wavelengths by NASA's Solar Dynamics Observatory on June 5, 2012. These transits could be detected by astronomers within a few hundred light-years, so long as they were in the plane of the ecliptic.
NASA/Goddard/SDO

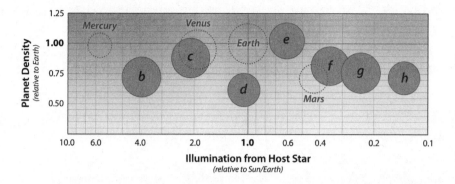

Properties of the seven known TRAPPIST-1 exoplanets (labeled b through h) compared to the terrestrial planets Mercury, Venus, Earth, and Mars, where the estimated bulk density of each (i.e., its composition—metallic or rocky or watery) is plotted in terms of the luminosity received from its host star. The relative sizes of the planets are indicated by the circles. The masses and densities of the TRAPPIST-1 planets were estimated from the slight variations in the timings of their orbits using extensive observations made by NASA's Spitzer and Kepler Space Telescopes, in combination with data from Hubble and a number of ground-based telescopes, and comparing these measurements with theoretical models. Estimates suggest the lower-density planets could have large quantities of water.
NASA/JPL/Spitzer Space Telescope

each other. So overall, astronomers are able to derive their densities and make educated guesses about compositions.

Uncertainties in these measurements will get beaten down in time. With luck we'll someday know whether they have oceans. With giant space telescopes we will try to subtract the raw starlight to isolate just the light reflected from the surfaces of the planets, to determine their colors and compositions so as to make sense of their atmospheres and determine whether they have clouds or continents or moons. I am extrapolating thirty years into the future, which seems fair; thirty years ago predates the discovery of exoplanets, so honestly, who knows what lies ahead?

The TRAPPIST-1 system bears a geometrical resemblance to Jupiter's Galilean satellites. And like those satellites, orbital resonances seem to have locked those planets together into an eternal pattern, linking their fates forever. The orbital periods in TRAPPIST-1 are in approximate integer ratios of 8/5, 5/3, 3/2, 3/2, 4/3, and 3/2, from innermost to outermost. Just as with the Galilean satellites, to migrate one, you have to migrate the whole chain, giving the pack of them a tremendous dynamical stability. It's analogous to how Pluto is dynamically coupled to Neptune, except that is the reason why Pluto is no longer a planet according to the IAU. Are the Earth-mass TRAPPIST-1 planets not planets?

An ultra-cool red dwarf has a slow burn, like a bed of hardwood coals. The prediction based on nuclear physics is that it will burn for trillions—with a *t*—of years, a hundred times longer than the present age of the Universe. Dynamical stability of the planets is ensured for hundreds of billions of years as well, the main risk being an interloper from outside the system (a *nomad planet*) coming in and colliding with one of their own. When you do manage to break a multi-resonant Laplace chain, the system can fall apart fast, so this would be a calamity of biblical proportions, but that might be likely only on a time scale of trillions of years.

If there is life on any of these worlds, it might have the opportunity to evolve for a thousand times as many iterations as life has evolved so far on Earth. If it hasn't started yet, there is certainly no hurry. Perhaps it will attain some kind of perfection when the Universe is done. Five billion years from now, looking up into *their* night sky, they will see our Sun expand into a red giant and then shed a beautiful nebula; it will then recede from memory, becoming another vanished star among the disappearing constellations. The TRAPPIST-1 system and all the other systems around the other red dwarf stars will live on, as one by one the brighter stars like the Sun flare up briefly and go dark. A hundred billion years from now, the aging, quieter Universe might still be pulsing with extraordinarily advanced life around these sturdy little stars.

The resonant chain of TRAPPIST-1 planets would make a very interesting calendar for their inhabitants and for some spectacular, never-repeating sights. The distance between the adjacent planets is a few times the Earth-Moon distance, and the planets themselves are a few times the size of the Moon, so at closest approach they'd appear in the sky like our full moon, although with different colors and markings. Sometimes they would be catching up, and other times falling behind, slewing under real and apparent motions as pairs of planets race around the star. Conjunctions of neighbor planets in the sky—"full moons," as it were—would happen every planet year—that is, every few weeks. And of course the planets themselves would have moons, and some might have rings. I think it is fair to say that while night skies on Earth are impressive, we have nothing on that.

But conditions around an ultra-cool M-dwarf have one major drawback. Look more carefully at the TRAPPIST-1 transit record, and you'll see that not only are there dimmings where a planet blocks the starlight, but there are bright spikes where the star emits a massive flare. The flash of visible light is what's measured, but associated with each flare would be all kinds of ionizing radiation exploding from unstable reactions inside of these barely burning stars. Life around TRAPPIST-1 and other systems, if it exists, might have to move underground or become oceanic to survive the radiation damage to living matter. Living underground is not so bad if you're a bacteria; Mars, Europa, Titan, and even Mercury[14] are thought to have regions below their surfaces that might be potentially habitable to organisms that survive on Earth. And living in an ocean is not a hardship. In any case, these stellar flares might also provide the right kind of spark for the origin of life—bursts of ionizing radiation that might trigger just the right proto-biochemical reaction.

I AM NOT sure we can quantify the probability of life until we discover it somewhere else, and am reminded of "the right method of philosophy" as postulated by the German philosopher Ludwig Wittgenstein, that "whereof one cannot speak, thereof one must be silent."[15] Still, alien life is what we think about and what we talk about, and to a large degree it's what planetary exploration is all about. There's certainly no harm in trying to frame the question. In making a deeper assessment of the problem, paleontologist Peter Ward and astronomer Don Brownlee[16] came to the conclusion twenty years ago in their book *Rare Earth* that complex life in the Universe is exceedingly rare, because too much has to go right.[17] And in 1950, the physicist Enrico Fermi remarked that with the hundreds of billions of stars in the Milky Way, we should have been contacted by now, unless complex life was exceedingly rare ("Fermi's paradox"). Maybe so.

Let's postulate that the origin of life is *deterministic*—that is, if you set up the initial conditions exactly, then life will happen. (If there was a spark of God involved, then I don't see why you couldn't cause life to take hold in the briny caverns of Ganymede and forgo plate tectonics and the Sun and the Moon and the rest.) If life's deterministic, then it comes down to how exacting these requirements must be. Let's assume it has some finite range, call it epsilon (the Greek letter ε, used in math to mean "something really tiny"), so if you make a planet within this ε-box of parameters, life is likely to happen. It's a dangerous game to play, because we are biased to think of our own experiences as normal, but let's play anyway.

Let's narrow the box to include only planets that have oceans and plate tectonics, what we tend to call "Earthlike." Maybe that's one in ten thousand (we do not know). Let's say that the planet also needs to have a large satellite that will drive tidal action of its early oceans, creating a richly interactive land-sea interface. Maybe that's one in ten (after all, we know of Earth-Moon and Pluto-Charon, so it's not uncommon). Let's also say that it needs one or more giant planets on more distant orbits that would serve as a filter to clear out late-bombarding asteroids and comets whose impacts would sterilize all life as soon as it had started. Here we might estimate one in ten thousand, given that our solar system appears to be unique among the thousands of planets discovered so far. So now we're at one in a billion. Let's even assume that life needs a neighboring planet like Mars, where life can develop first while the long-term habitable planet (Earth) cools down from its superheated state. I'd guess, one in ten. Let's also say there have to be enough asteroids and comets around (but not too many!) so that one of them can end up knocking life off the Mars-analogous planet onto the Earth-analogous one. Maybe that's one in ten, so we're down to only about two such planets in the Milky Way. The central star needs to be relatively stable and have an expected lifetime of several billion years; maybe we require it to be a sunlike star, so now let's say one in a hundred. Now there's fewer than one in the galaxy. Maybe it also requires a late delivery of biogenic molecules to the planet from somewhere else in the system, like phosophorous or carbonaceous materials, by infalling asteroids. I think you get that for free when you invoke ballistic panspermia, but let's add another factor of 10 just to be sure. And then let's say it has to be in the right part of the galaxy not to get sterilized by stellar activity and gamma ray bursts, maybe one in a hundred. Last, for good measure, let's say you have to have a late K/T-type impact in there somewhere, mass extinctions that would allow the "underdog" species (in our case, mammals) to crawl out of their holes and take over the planet, so, one in a hundred. Now we're at one in a billion trillion. That value of ε would imply there are at least a hundred thousand very close copies of the Earth in the Universe—if the darts are always thrown at random!

These sorts of arguments—if you throw a hundred billion trillion darts at the dartboard, then you're likely to hit the bull's-eye a hundred thousand times—have limited utility in terms of predicting anything, but serve as a framework. For example, what if planets are not each unique, and that instead of a hundred billion trillion random experiments, there are a hundred trillion close copies of only a few billion blueprints for planet formation? Then the odds of finding anything inside of the ε-box can be near zero. If planet formation is like making raindrops, then you could make an infinite number and they would all have some identifiable variation—salinity, diameter, temperature—but they'd all be raindrops.

Counter to this is the evidence for substantial planetary diversity; I think inefficient accretion maximizes planetary diversity, ensuring that instead of just raindrops we have ice balls, rock balls, water worlds, metal worlds, and the potential for everything in between. Maybe no matter how small an ε you define, this diversity will guarantee that somewhere, sometime, you'll hit the bull's-eye and make life.

What if humans truly *are* precious, a singular spark of advanced consciousness on the verge of moving out into the cosmos? I think of the many times I have started a fire and the flame went out when I wasn't paying attention, followed by half an hour of struggling in the cold to get it going, having spent all my dry twigs. Is that us? If we are in fact alone in our galaxy and if we don't care about *now*—if we let this good house fall apart—then all of our best creations, the sum of civilizations past and present, will be lost, irrecoverable, and arguably meaningless, the Universe going on as though all of this never was.

If complex sentient organisms like ourselves *are* common

throughout the galaxy—cozy little campfires here and there—and we carelessly let our own little fire go out, then the record of what we've done here could be a mystery for future alien geologists to solve, the squiggles on a plot of carbon-oxygen isotopes, and core samples of seafloor sediments incorporating synthetic molecules that point to the existence of a long-ago industrialized species. The Holocene-Anthropocene transition is being laid down now, with new sedimentary and volcanic rocks being constructed around swarms of plastic bits, and the dissolution of carbonates and corals by a warmer, more acidic ocean and the seasonal runoff from bigger storms. Millions of years from now Earth could provide a textbook example, maybe sadly common throughout the galaxy, of a creative and advanced civilization that fell from the edge just as they had opened their minds to *see*.

The galaxy as it appeared from Wellington, New Zealand, on October 25, 2013.
Photograph by Andrew Xu (CC BY-SA 2.0)

CONCLUSION

I N ONE OF THE MORE reckless gambits of planetary exploration, Christopher Columbus risked his crew's life on an unsafe
bet, not that the Earth was round, but that the Earth was *small*.
India was already an overland trading partner for spices and
other treasures; Columbus proposed an overseas voyage that would
demonstrate a better, faster, cheaper[1] way of getting there. Queen
Isabella of Castile was quite familiar with the idea that the Earth
is round; what Columbus sold her on was his confidence that the
Earth is only 5,000 kilometers in diameter. That would mean that
by sailing west, India would be only 2,000 kilometers away. I suspect Columbus was well aware the Earth is more than twice as
big, a value that had been established since ancient times, and that
he intentionally underbid the cost, complexity, and duration of the
voyage to make it happen. This would not be the first time or the
last that a mission of exploration was sold by a reckless optimist.

Except for the intervention of an unexpected landmass that
would later be called America, and specifically a little island in
the Bahamas, his adventure would have been a disaster. Columbus
didn't discover America; he got there despite his intentions. (Besides, the Vikings had arrived hundreds of years before, although
their voyage didn't require believing the world was round; they
island-hopped from Iceland to Greenland to Newfoundland.) The
Americas were already populated by tens of millions of natives
who were there all along, having spanned both North and South
America after crossing the land bridge from Asia tens of thousands
of years before. Maybe some of their astronomers also had figured

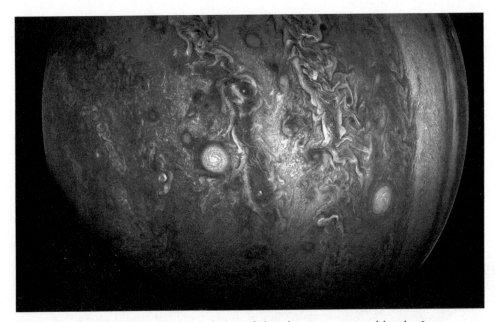

Cloud tops of ancient Jupiter, father of the planets, as imaged by the Juno mission. The color images are breathtaking. Since the Juno camera was not a science instrument per se, the largely unexplored and unprocessed raw data have been in the public domain since the start of science operations. Creative scientists and artists (amateurs and professionals) have processed them into stunning images, including many wonderful flyover movies, a number of them set to music (e.g., Gustav Holst, Lukas Ligeti) that really puts you there.
NASA/JPL-Caltech/SwRI/MSSS

out that the Earth was round—they too undoubtedly made the observation that the Earth casts a curved shadow on the Moon—but they certainly weren't expecting what was to come from beyond the waters.

Columbus's first voyage took more than two months. Going to the Moon takes only three days; it's like crossing the English Channel in a bathtub in good weather. NASA did it several times, and the astronauts who journeyed to the Moon and back all had good fortune. Going to Mars is like crossing the Atlantic on a one-way voyage that takes a hundred times as long, and where one or two storms are expected; you'd die for sure if you tried it in a bathtub. We can certainly build bigger ships, but an immediate challenge for astronauts voyaging to Mars is the intensive radiation damage their bodies will experience from the torrent of high-energy particles from the Sun, with a major event becoming probable, and cosmic rays from the monstrous explosions throughout the galaxy.

High doses of radiation are lethal to humans and other organisms, breaking down DNA faster than it can be repaired. Radiation also damages computer hardware and memory—no small risk when replacement is impossible. On trips to the Moon, astronauts accepted no small risk. On a voyage to Mars, lasting nine months on an optimal trajectory, radiation is not a risk but a death sentence unless the vessel is massively shielded. This shielding could be a thick metal hull, or water tanks and supplies, or even regolith scooped from an opportune asteroid—but you have to design it, build it, launch it, and give it propulsion. Despite what you see in the movies, we're not close to going to Mars in human form.

While Mars is perhaps more captivating to futurists because of its spectacular vistas, its wee bit of atmosphere, and the idea that you can live off the land, it's much easier to establish a settlement on the Moon, at least in the near future. For one thing, the radiation problem is readily solved: after a three-day voyage that you can handle, timing it to avoid a solar flare, you land and move into a bunker in which you sleep and hang out during radiation events, and from where you can do limited extravehicular activities. The only thing it lacks is water, food, and air—those you have to make on your own.

One candidate for a lunar base is the high rim of Shackleton,[2] a 20-kilometer crater near the Moon's South Pole. The highest rim is in near-continual sunlight, and it has an almost constant line of sight with Earth for communications. A few kilometers down the crater wall, you encounter a pitch-black crater floor, a place of such darkness that it has not seen a direct photon from the Sun for billions of years. You would not want to fall down there. In this frigid wasteland, as cold as Pluto, massive layers of ices and hydrocarbons have accumulated, molecule by molecule, into a hoard of treasure. The main thing lacking on the Moon is water, and here it is. Solar energy from up on the rim can be used to power the electrolysis of H_2O, converting it into Hs and Os so you can breathe. Rocket fuels, chemical fertilizers, plastics—all kinds of things can be manufactured depending on what's found down there and in the surface regolith. You can manufacture a kind of cement or mortar, and extrude it to print 3D structures and habitats.

Do I worry that we might trash the Moon? The idea mortifies me. It's the pity of desert places, that when humans come around with an industrial purpose or with the bored but innocent goal of roaming around and making our mark—the land becomes degraded and ugly. I am a strong advocate of human bases on the Moon, and even mining on the Moon to build colonies, and using the Moon as a stepping-stone to planets such as Mars, but we need to be reasonable. The Moon is irreplaceably special to our species. When I was a kid watching the first human come down the ladder, I was told that those bootprints would be there for thousands and thousands of years, since there is no rain or wind in the Sea of Tranquility. Will they survive space tourism?

SO WE COME back to where we started. The Moon is a symbol of religion and our psyche, an object of the most serene beauty, yet also a scientific object to be explored and a breathtaking place to visit. As for the resources it offers, we see on Earth the disaster of mining when it disrespects the landscape, and the unnecessary loss of the last of the wild places. One could limit colonization to the lunar nearside, for example. But it's a mistake to pitch one reality against another, because all of it is what it is to be human.

It is wondrous to behold the red of sunset, knowing it is caused by the wavelength-dependent scattering of light. It makes diamond earrings more beautiful to know that their stones had erupted in kimberlite pipes from the base of the crust. Knowledge of the natural world—what I am sitting on, where I am at, what I am seeing— brings you closer to creation. What color is the best? I cannot tell you. Is there a God? I cannot convince you. But which mineral is harder? What is the wind speed over those clouds? Is the Andromeda galaxy going to collide with the Milky Way? These questions are answerable, and ignite our passion to get to the bottom of things— the human emotion of *curiosity,* the quality of spirit embodied by our hands and fingers, our eyes and ears, and our brains.

Within the dominion of *that which is the case,* science has enormous practical significance. If basic Newtonian gravity or the principle of solid friction were someday proved false, you might float up from the floor or your house might slide into the sea. That's the stuff of dreams, where the laws of nature no longer hold—and while nobody's to say that life's *not* but a dream, the world will do what it does and will be what it is, whether or not there are scientists. If you jump from a high cliff you will die, whether or not you believe in the principles of gravity and inertia.

Science is replete with theories that have no immediate significance. That the Moon was created in a giant impact or that its farside highlands were born in a giant splat—these are interesting to consider, but it does not directly matter whether they are right or wrong. But quite possibly they matter indirectly, through the principles of deduction. If *A* is true, then *B,* leading us to change the questions, and learn from things that seem to be minutiae or that strike you as odd, that things don't quite add up. Sometimes when you are stuck, all that's required is a change of view, a good walk. A mission to Titan or to Venus might be the most important thing we do in addressing climate change on Earth.

Doing science is kind of like showing up in a foreign land and watching kids on a field play a game. You have never seen this game and don't know the rules. You could ask, but you don't know the language. You want to figure out the rules so that you can play. Some of the rules are immediately clear: two teams are trying to get a ball into each other's net. That's like the law of gravity: the most basic rule that seems clear from abundant observations, without which there would be no game. Other rules you won't figure out until you've watched dozens of games: they use their feet, and sometimes their heads, but never their hands except for one player on each team. And then there are behaviors and patterns that have no rules, and it is folly to figure them out until you get right in there and start playing and make mistakes and learn.

Another analogy is that scientists make paintings and hang them up to dry. They are arranged in a community effort that puts some on display and moves others to the basement. The process is not always fair—some of our best works are down there! On occasion, some of the most popular paintings need to be taken down, displaced by something deemed better. You can go down to the basement at any time—especially in this widely connected and archived computer age—but the stuff on display, articles that win their way into the pages of books and high-profile journals, define the conversation. Sometimes something from the basement is brought back upstairs. Other times, as Harold Jeffreys advised, we have a spring-cleaning and bonfire.

If the medium and the canvas are objective reality, then the brush, I think, is logic and mathematics. It is no coincidence that mankind's earliest astronomy overlapped with the invention of geometry, the way that modern geophysics began with the determination, thousands of years ago, that the Earth is a sphere, and that the Moon is thirty Earth diameters away, and one-quarter of its size. Geometry applied to the measurement of space yields indemonstrable truths. With time, using better measurements and more refined geometrical principles, astronomers would determine the distances to the farther planets, and their orbits, and the distances to stars, and within the last hundred years, the distances to galaxies, and the size and the inflation of the Universe.

Is scientific reality invented or is it discovered? However you feel about science and its role for modern humans, there is a powerful empirical argument that science *works*. Hundred-ton airplanes glide through the sky, full of passengers eating breakfast. A video call from your sister streams to your wristwatch. Ten-mile-long bridges and tunnels span a megalopolis, and its hundred-story buildings sway through the typhoon without damage. A new medicine eradicates a disease. These and other marvels are based on the application of the scientific method to learn the underlying principles (induction), and then the application of those principles, using mathematics, to design systems that function predictably (deduction). And all of this enables new technologies—the telescopes, the lasers in the lab, the deep-space missions—that enable further exploration and analysis at the molecular level, or at conditions deep inside the Earth or inside the stars, and out in deep space and around other planets, and the fabrication of human habitats in space. The more we know, the farther we go and the more we learn.

SAGES AND PRIESTS of mystical antiquity gave each planet symbols, and although it's not especially convenient to use them—they aren't included with standard computer fonts—astronomers still embrace them. In academic journals you see the radius of Saturn written $R_\hbar$ and the mass of Mercury written $m_\male$ and so on. Science is hanging on to a little of the symbolic magic, *just in case*. We also incant the names of planets every week, which in English is a curious mix of associated deities: the Sun, the Moon, then the Norse gods for Mars (Tiw), Mercury (Odin), Jupiter (Thor), and Venus (Freya), then on to Saturn, ruler of the bountiful golden age.

Saturn acquired a deeper meaning for me when I saw it in my friend's backyard telescope one summer night. I was about five and it was difficult to reach the eyepiece. His telescope was inexpensive and jittery, but he was proud and happy and had set up a box for me to stand on. There in his little eyepiece (have a look!) was Saturn, *really there*, not at all like a picture. That was sunlight coming into your eyes. It had traveled a billion miles from the Sun to Saturn, taking an hour to get there, and reflected back toward

Hubble Space Telescope image near opposition, in 2018, when the Sun, Earth, and Saturn formed an almost straight line. Saturn's rings were near maximum tilt as viewed from Earth.
NASA, ESA, A. Simon (GSFC) and the OPAL Team, and J. DePasquale (STScI)
(CC by 4.0)

Earth, another hour, and a small fraction of those photons gathered into formation to create the image of a ringed world in your eye.

The yellow-salmon disk with its pale stripes and golden rings raced from the right to the lower left, as the telescope, fixed to the rotating Earth, slewed across the highly magnified view. I tried to nudge it back into the center, then lost it, and my friend took another ten minutes finding it. It didn't matter. Reality had drawn into sharp focus. The pale star in the backyard sky was no longer a twinkle—it became a floating jewel.

EPILOGUE

AS SCIENCE PEERS FARTHER AND deeper, coming closer to the origins of things, we are also pushing at the limits of our planet. Today humans and livestock account for 96 percent of the mammalian biomass (an age of hamburgers) and our existence is amplified by powerful, globally disruptive technologies and wasteful practices. At the same time, we have discovered a universe so vast and intricate that we feel sparse and alone, staring at the Hubble Deep Field with its thousands of galaxies, each with its billions of planets. We've painted ourselves into a corner, yet we're lost.

To break out of our celestial shell—stars painted on the canopy— and step out into the bigger picture where real explanations exist, we need the vantage of planetary exploration, including the Earth and our place in it. That's why I've written this book. I think we ought to double our exploration budgets to the levels of the 1960s to learn more about the Earth and Moon, Mars, Venus, Titan, Pluto— all of them—and start designing probes to nearby planetary systems; start soon and they might get there in our great-grandkids' lifetimes. Of course, some would say the ways of the 1960s are over, good riddance, and anyhow now we have climate change on the front burner; outer space must wait. But there is no outer space. We are in space, and our climate is the boundary of that.

If we are interested in practical solutions to global problems, we might begin with the fact that *Homo sapiens* tend to aggregate into colonies or "nations" that compete for the status of combative readiness and technological prowess. That has been true since before the time of Thales, and has depended increasingly on scientific capability: better catapults, better warships, better lasers. But warfare is a visionless industry, a destructive application of the greatest science and technology. The Apollo program, now *that* was visionary—yet tied to the very same industry! What made it so? It was built on the same rockets, and relied on the same factories and expertise, and many of the same systems and personnel. There's nothing wrong with rockets; it's what you do with them.

Apollo was a muscular display of superpower derring-do, far more impressive than rolling out a new line of fighter jets or nuclear submarines, and costing much less. The touchdown of the *Eagle*, fifty years ago this past summer, may be the most longstanding singular achievement of that century, and unlike the warfare of that era it brought us together as a species, however briefly. And apart from the discoveries it made, Apollo paid for itself many times over by raising the value of American capital, proving that money in and of itself is not a thing. As with rockets, it's what you do with it.

Here we are, on a mote inside of a fractal inside of a fractal, on an incomprehensible scale. Me, you, the kids, the polar bears and turtle reefs, anything we do, whatever we create—how can any of it matter? Sometimes we get a nudge, a hunch we have to follow, that leads us out. For me, that hunch is that the ecological crisis is a wake-up call that any sapient species will ultimately face, a rite of passage. Winning becomes too easy, and a transformation of intention is required. The way we define success has to shift, from our old habit of expansion and dominion—using every new tool to gain more leverage—to qualities that are fundamentally mammalian: the nurturing of offspring and living in alliance.

Nations compete intensely in the race to the Moon, but this race has also highlighted our most basic human tendency for cooperation. Olympic athletes will help each other fix a flat tire on the way to the competition. The venerable Soyuz spacecraft, originally intended to ferry Russian cosmonauts to the Moon, has been used to fly dozens of NASA astronauts to and from the International Space Station, NASA's only choice after the grounding of the shuttle. Sharing data is also common; NASA's highest-resolution images of the Moon plus all kinds of other geoscientific data are available to Chinese mission planners so they can prepare for their own crewed landings, which will happen sooner than you think (and, boy, will that be news). Then there's the Apollo-Soyuz Test Project that ended the Apollo era and thawed the cold war. Crews launched separately from Baikonur Cosmodrome and Kennedy Space Center docked in low-Earth orbit and spent two days working closely together, the culmination of years of unprecedented planning and the sharing of sensitive technological information. How could this be, when just a decade before, men from those very nations almost sent us into thermonuclear war? A human space mission, even in low-Earth orbit, is a trusting collaboration involving engineers, astronauts, and ground crews, and supported by political and military leaders, realists who happen to be futurists, who recognize the human potential to cooperate in outer space.

Some of us are waiting for aliens or deities to beam us up. The answer may indeed come from space, but if so, maybe more likely in the form of asteroids, the small stray bodies from which planet formation began, like the big one that wiped out the dinosaurs so that mammals could rise up 65 million years ago. Sooner or later, if left to their own devices, one of these asteroids will again bring devastation, and I think that's lucky news. On Friday the 13th in April 2029, the 370-meter diameter asteroid Apophis (2004 MN4, 50 million tons, god of chaos, enemy of light) will fly closer to the Earth than the orbiting communication satellites. It will be a backyard spectacle, this naked-eye star moving visibly against the night sky, notorious for having had a substantial probability of hitting us when it was discovered. A collision is ruled out; instead, Earth's gravity will bend its orbit on an arc and send it in a new direction. What great nation will send a reconnaissance mission to Apophis? Will a group of university students beat them to it with a "cubesat" flyby? Who will send back high-definition footage of the irregular object as it approaches, looms over Earth in full-frame for about an hour, and heads out? Maybe somebody will be able to land on it, prior to the encounter, and use seismometers to listen to it creaking in response to Earth's tidal field, like the hull of a ship caught in sea ice.

The nations and universities and private explorers who rally around NEOs will likely be the ones who participate in human bases on the Moon, and in cislunar space. As for the farther reaches, Japan's upcoming Martian Moons Explorer may be a pre-lude to international outposts on Phobos and Deimos, where we would make use of the same technologies developed for NEOs. As for the raw economic lure of asteroids, Apophis, a common S-type, is worth $10 billion in rare metals if you could conceivably get to them. More exciting would be a primitive carbonaceous asteroid, especially if you could maneuver it into a reliable retrograde or-bit around the Moon: for in addition to silicates and rare metals it would have water and other volatiles that would be needed for mining and propulsion, and to enable closed-system life support and agriculture, and the raw materials for fabricating massively shielded structures.

But asteroid resources are as much of a carrot as the hazard is a stick. We are the stubborn mule. So let the hazardous asteroids knock, and bring out our better nature. If one of them leads to a space race to divert it, the response will be a coordinated activity, like in 1986 when spacecraft from Europe, Russia, and Japan flew by Halley, but with a lot more muscle. As for making money from asteroid materials, that future has not yet arrived, and won't until we develop an economy in space. But this whole idea, of once-hazardous asteroids captured as international resource outposts, while the Earth is returned to its garden state—these are no more chapters of science fiction than reading a book on a tablet while gazing out of an airplane window would have seemed to a cowboy of the 1850s.

Astronaut carrying experiment packages on July 20, 1969. Apollo Image Atlas AS11-50-5944 (70mm Hasselblad).
NASA/LPI

ACKNOWLEDGMENTS

I AM GRATEFUL FOR HAVING HAD gifted teachers, and students whom it's been a privilege to teach, and colleagues who have shared their ideas and discoveries, whose work I hope to have fairly represented. I'm also grateful for the artists and photographers who have contributed their talents to these pages, and the spaceflight engineers and mission specialists and astronomers and astronauts who have brought home pictures and other data from planets near and far; one image can be the culmination of decades of effort and adventure. I'd also like to thank my editor, Geoffrey Shandler, who cleared out the debris of manuscript accretion, especially when the paragraphs were crashing into the Sun, or being swept up by other chapters or scattered to the outer regions. And to my dear parents, Gunnar and Tulla, without whose guidance I might have been writing a book about John Donne's use of metaphor instead of grappling with the infinite. And to my wife and kids, who have been supportive and kind while I wrote my big book about planets: I love you.

GLOSSARY

absorption line
A feature in a spectrum, caused
by the absorption of a particular
wavelength of light by the
presence of, for example, certain
molecular bonds or electron orbital
transitions. The depth of the
absorption indicates the abundance
of whatever species is present.

accretion
Growth of discrete bodies by
accumulation of mass from the
planetary disk.

achondritic
An asteroid or meteorite of evolved
or once-melted composition.

anorthosite
A cumulate rock made mostly
of feldspar that comprises the
highlands crust of the Moon.

aphelion
The point in a planet's orbit when
it is most distant from the Sun
(compare *perihelion*).

Apollo
The crewed space missions that
brought twelve astronauts safely to
the surface of the Moon and back.
Also, a family of asteroids whose
orbits come close to Earth.

asteroid
A small rocky body originating in
the terrestrial planetary region.

astronomical unit (AU)
The average distance from the
Earth to the Sun, 149.6 million
kilometers.

basalt
The most common terrestrial
surface rock, made of solidified
silicate lava.

carbon cycle
In plate tectonics, the movement
of carbon from the atmosphere,
into the ocean, to the seafloor
as carbonates, into the upper
mantle, and back out of volcanic
vents.

chondritic
An asteroid or meteorite of
composition undifferentiated from
the condensable solid component
of the original protoplanetary
nebula (also *solar composition*).

comet
A small, ice-rich planetary body
orbiting the Sun, which originated
far beyond the terrestrial planet
region.

cometesimal
An ice-dominated planetesimal.

co-rotation radius
Orbital distance at which the period
of the satellite orbiting the planet
equals the rotation period of the
planet.

cryosphere
The exterior frozen solid shell of a
water-rich planet or satellite.

C-type
Dark reddish asteroids from the
outer Main Belt, parent bodies
of the carbonaceous chondrite
meteorites.

delta-v
The change in velocity required to get somewhere in the solar system; or, the amount of velocity change to a spacecraft that is attainable by a given rocket motor.

density
The mass of something, per unit volume. Iron is almost three times as dense as rock, which is three times as dense as water.

differentiation
The process by which a planet forms a core and mantle and other layers, usually by melting and segregating gravitationally.

embryo
A planet that is rapidly growing and at risk for being accreted.

extinct radionuclide
A radionuclide such as ^{26}Al or ^{60}Fe with a decay half-life shorter than the early evolution of the solar system, which may have been abundant but is now gone.

gardening
The process by which small impacts overturn and reduce to powder the regolith on an asteroid.

geometric series
A series of numbers where one is a constant factor times the preceding—e.g., 1, 3, 9, 27, . . .

grand tack
The theory that Jupiter originated where the Main Belt is today, at around 3 AU, then migrated in toward where Mars is today, at around 1.5 AU, and then when Saturn was formed, the pair of them migrated back out to their present locations at 5 and 10 AU.

gravity assist
When a spacecraft makes use of a planet orbiting a star, or a massive satellite orbiting a planet, to speed up or slow down or otherwise modify its own orbit without any expenditure of fuel.

half-life
The time for half the atoms to decay in a radioactive sample.

hit-and-run collision
A common type of planetary collision in which the impactor and target are of comparable size and do great damage to each other (especially the smaller) during the event, but do not accrete.

hydrocarbon
A compound made of C and H, e.g., methane (CH_4), and ethane (C_2H_6), found in reducing conditions.

ISRU
In Situ Resource Utilization, that is, using what's available for propulsion, life support, habitation, etc.

KREEP
Acronym for *p*otassium, *r*are-*e*arth *e*lements, and *p*hosphorus, the residue of a primitive magma ocean that is concentrated in incompatible elements including uranium and thorium. Found in abundance on the lunar nearside, and rarely on the farside.

low-Earth orbit (LEO)
The easiest orbits to attain, a few hundred kilometers above the atmosphere of Earth. This is where the International Space Station orbits.

Main Belt
The debris disk between Mars and Jupiter where most asteroids are found. With a total of 5 percent the mass of the Moon, half of the Main Belt is contained in the four largest asteroids, Ceres, Vesta, Pallas, and Hygiea.

mascon
A gravity anomaly (mass concentration) in the crust of the Moon or a planet, which can be caused by dense mantle rock flowing into the volume displaced by a large impact crater.

meteorite
A fragment of planetary material that lands on Earth.

meteoroid
A geologic mass in space that is smaller than an asteroid—that is, smaller than about 50 meters.

near-Earth object (NEO)
An asteroid or comet that has perihelion distance inside 1.3 AU.

oligarch
Any of the final planetary embryos after orderly growth, at the beginning of the late stage of giant impacts. Theia was an oligarch.

olivine
The most common mineral in Earth's mantle, with a composition ranging from Mg_2SiO_4 to Fe_2SiO_4.

orbital resonance
When two orbiting bodies are periodic. Usually we mean that the orbital periods are in a ratio of two small integers, but in a *secular resonance* the precessions of the orbits are synchronous.

perihelion
The point in a planet's orbit when it is closest to the Sun. Earth's orbit is slightly eccentric, so on January 3 it is about 3 percent closer than at aphelion.

photosynthesis
The formation of carbohydrates from carbon dioxide and a source of hydrogen in chlorophyll-containing cells exposed to light.

planetesimal
A primordial planetary body large enough to be gravitationally bound.

primary accretion
The formation of the first macroscopic condensates in the nebula, from centimeter-scale lumps to kilometer-scale planetesimals.

protoplanetary nebula
The cloud of gas and dust from which the Sun and the solar system formed, sometimes called just "the nebula" or "the disk." Also, any such planet-forming region elsewhere in space.

reduced
Conditions that preclude the formation of oxides, such as the hydrogen-rich conditions of the early nebula, or the atmospheres of the giant planets.

regolith
Loose granular material on the surface of an airless body.

resolution
The finest scale at which you can measure something in the data. A typical movie has a time resolution of 1/30 of a second, for example; image resolution is typically a few pixels.

rheology
How a material flows or deforms in response to a stress condition, and conversely, how stress builds up inside a body when it is deformed.

Roche limit
The orbital distance inside of which an accreting satellite is always under some kind of surface shear stress that can pull it apart. Because satellites aren't liquids, the tidal disruption radius is closer than the Roche limit.

rubble pile
A geologic mass whose chief binding force is self-gravitation, but whose central pressure is far lower than the strength of rock. A rubble pile cannot hold on to its pieces if it is spun up to rapid rotation.

scaling
A physics-based mathematical transformation allowing laboratory results to be extrapolated from one domain to another—e.g., to a much vaster scale, or to much longer time.

small planetary body
A comet, asteroid, or small satellite; any geologic body where gravity is extremely low.

solar pressure
The density of the linear momentum of the Sun's electromagnetic field, which applies a pressure that can be felt by small particles and spacecraft.

space weathering
The process of surface reddening and darkening that occurs as minerals and metallic inclusions are exposed to the high-energy radioactive bombardment of space.

S-type
The most common type of asteroid in near-Earth space; parent bodies of the ordinary chondrite meteorites.

supernova
A massive star that has reached the end of the fusion process and undergoes core collapse and eruption.

terrestrial planet
Any planetary body or satellite whose dominant processes are closely analogous to principal processes on Earth.

tidal evolution
The migration of a satellite's orbit due to transfer of angular momentum from the planet's rotation to the satellite.

triple point
The temperature and pressure where the standard phases of pure water coexist: liquid, vapor, and solid. With its atmosphere, Earth's surface environment is around the triple point.

Trojan
A body that orbits at one of the stable Lagrange points, L_4 or L_5. Also known as a *co-orbital*.

NOTES

A SHORT LIST OF PLANETS AND MOONS

1. Data for planets and satellites from https://ssd.jpl.nasa.gov. Data for the Moon and planetary orbits from https://nssdc.gsfc.nasa.gov/planetary /factsheet. Data for Haumea from D. L. Rabinowitz et al., "Photometric Observations Constraining the Size, Shape, and Albedo of 2003 EL61, a Rapidly Rotating, Pluto-Sized Object in the Kuiper Belt." *Astrophysical Journal* 639 (2006): 1238–51, and D. Ragozzine and M. E. Brown, "Orbits and Masses of the Satellites of the Dwarf Planet Haumea (2003 EL61)," *Astronomical Journal* 137 (2009): 4766–76.

INTRODUCTION

1. Opposition is when the Sun and Moon are on opposite sides of the Earth. The Moon is then completely full. Most oppositions are not perfectly aligned, so the Moon stays outside the Earth's direct shadow; it is then a very bright full moon because of the so-called opposition effect, where light is reflected from the Sun to the Moon's powdery surface and then directly back at you, similar to a cat's eyes in headlights. When oppositions are perfect, it is a lunar eclipse, and the Moon passes through the Earth's shadow for a few hours.

2. In order of mutual generation, wood, fire, earth, metal, water; in order of mutual overcoming, wood, earth, water, fire, metal. These cycles may be a pattern of terrestrial planet building, if *wood* represents organic matter.

3. It's safe to stare at the Sun if you buy a pair of hobbyist binoculars with solar filters for around $40; your science teacher may have a pair.

4. It does to accountants, though. The businessman and photography pioneer George Eastman, founder of Kodak, advocated the 13-month, 28-day International Fixed Calendar developed by American railway adviser Moses Cotsworth, and recommended it to other businessmen of the 1920s. His company continued using it to define 13 pay periods until 1989, with 28 days per month and an extra day on the last one. (My former employer had eighteen 20-day pay periods plus a bonus period; you might as well go full Babylonian at that point.) Cotsworth

originally envisioned a thirteenth month called Sol in between June and July. Of course, the Moon would still have done whatever she wanted, being offset from this cadence, a synodic month (full moon to full moon) taking 29.5 days.

5. John Keats, "Ode on a Grecian Urn" (1819).

6. One rotation from noon to noon, that is.

7. The Hindu week also uses planets, but in order they are the Sun, Mars, Jupiter, Saturn, Moon, Mercury, Venus. The cycle of planets in the Chinese calendar is Sun, Moon, Mars, Mercury, Jupiter, Venus, and Saturn. I suppose there's something to that, how a culture decides to allocate planets to days.

8. The day is one rotation of the Earth, but in one day, the Earth goes 1/365th the way around the Sun, so a solar day is that tiny fraction longer. Likewise, there is the synodic month and the sidereal month, the synodic month being the period of the Moon orbiting the Earth, relative to space, and the sidereal month being the period between one full moon and the next—where the arrow pointing to the Sun has advanced by about 1/12 of a year in one lunar orbit. So a sidereal month is 27.3 days, and a synodic month, the time between full moons that is the basis for the lunar calendar, is 29.5 days.

9. An astronomical fortnight is half a moon, 14.77 days. The word means "fourteen nights."

10. It's important to have champions of probably wrong ideas, and that is the significance of tenure. For example, Avi Loeb, chair of Harvard's Astronomy Department, is championing the hypothesis that the object 1I/'Oumuamua from beyond the solar system is actually a fragment of an interstellar spaceship or light-sail material. It's important that he do so. One can then follow through on the consequences to find the reductio ad absurdum, thereby eliminating the hypothesis, or not, in which case the hypothesis gets a bit stronger with every test it survives. Either that, or it morphs into an unfalsifiable claim, akin to Bruno's theories.

11. Galileo Galilei—"A Gentleman of Florence," as he called himself on the title page of his most famous work, *Sidereus Nuncius* (1609)— described in vivid terms how his 30-power telescope brought the Moon closer: "It is a most beautiful and delightful sight to behold the body of the Moon . . . about thirty times larger."

12. Genesis 1:9.

13. Edward Tarbuck and Frederick Lutgens, *Earth Science* (Columbus, OH: C. E. Merrill Publishing Company, 1985).

14. As we'll see, this was the popular paradigm from the late 1800s until the mid-1920s, and it resonated well into the 1960s. At the time it was called the planetesimal hypothesis to distinguish it from the Kant-

Laplace nebula hypothesis. Today, by *planetesimal* we mean something completely different, so I call it the stellar collision hypothesis.

15. Russia had a phenomenally successful series of Venus landers in the 1970s and early 1980s. Venera 7, in 1970, was the first spacecraft to land on another planet besides the Moon. It would be followed by a half dozen other landers that would study the atmosphere and soil chemistry and return detailed close-ups.

16. It was the time of Madonna's first album, an age of ditto machines, small cylindrical printing presses that you filled up with lighter fluid (or something like it) and rotated the handle to spin out copies from a ditto master that you made the night before by typing really hard (using a mechanical typewriter) to get the ink to transfer evenly on twenty-two sheets of paper. You'd roll the thing with a good rhythm, and your friend would get her coffee and another would head out to the smoking shed, and the rest were already in class because you were late.

17. Marine biologist John Menke loves to teach and has inspired a number of his own students to become teachers, and in this case, teachers to become scientists.

18. It being a small school, I also taught sophomore English, where I grew to believe that English teachers deserve higher pay, or else smaller classes, for all the grading they do, and all the classroom interaction.

19. The equations of calculus are simple; call the velocity v, then the acceleration is written as dv/dt, where t is time and d is the differential. So acceleration is change in velocity in a change in time, meters per second, per second.

20. A PhD candidate is a student who has passed their qualifying exam, is finished with classes, and is working to become the world's expert in some field.

21. Launches of astronauts in the Soyuz capsule from the Baikonur Cosmodrome were even more routine, but not televised.

22. Christa McAuliffe, a teacher from New Hampshire, was selected from more than 11,000 applicants to be the first teacher in space.

23. About 1 launch vehicle failure in 25 launches was expected, so it was not particularly astonishing. Contributing to the investigation, physicist Richard Feynman concluded that the O-ring design was inadequate for the weather conditions that were anticipated; this led to the rupture of the *Challenger* fuel tank. And it was a flawed design to begin with, to have the nose cone in front of the human spacecraft. Ice chips were shed from the rocket fairing and cracked the tiles on a later mission, in 2003, causing *Columbia* to disintegrate upon reentry, killing all seven onboard.

24. It also created an interesting geopolitical situation where we relied on the Russians to take our astronauts to space. The Soyuz, which

continues to maintain the highest success rate of any large launch vehicle, has never undergone any major design modifications. Don't mess with a good thing. The United States should have kept flying the Saturn V.

25. The technology in the Galileo mission was frozen at 1970s technology, because every mission has a "tech freeze" where only space-qualified components are included. So, for example, Galileo actually used a magnetic tape on a reel to store data. As data sectors went bad, the engineers would upload code to skip over the bad sector.

26. It was powered by the radioactive decay of 50 pounds of plutonium dioxide. Because plutonium dioxide has a half-life of 87 years, the power pack was unaffected by the delay.

27. By the early eighties, when software was finalized for Galileo, there was no such thing as image compression. Engineers studying newfangled wavelet-based compression technologies were motivated by the Jupiter data problem, and came up with algorithms that had to be uploaded to the spacecraft computer system and comprehensively tested onboard. This was at no small risk to the already taxed mission, which relied on a radiation-hardened mission computer with the capability of an Apple II computer of that era.

28. If you have an idea for how to spend an hour or even a whole morning with a class of students, do a good thing and step forward at your local school. Teachers are grateful. And if you have some quality equipment you can donate to a small school, such as microscopes and telescopes, that's good too!

29. William Wordsworth was the early nineteenth-century English poet whose method of wandering I appreciated, and James Hutton was the Scottish geologist of the middle 1700s, originally an agrarian farmer, who introduced sedimentology and the notion of deep geologic time to Western thought.

30. A wash is a desert creek that can be as wide as a river that flows only in the monsoon rains.

31. The LSST (Large Synoptic Survey Telescope) primary mirror is 17 tons and 8.5 meters in diameter; it was cast on a huge rotating furnace whose spin gave it an approximately parabolic shape from the centrifugal forces. Archimedes would be impressed.

32. A land mammal's body is about a meter, because it must be held together in the planet's gravity, and that requires 10 kilograms of skin, for each of us, and many pounds of deeper fascia and connective tissue, supported vertically by a skeleton and long, strong muscles. Bigger bodies require much greater structural integrity and expense of energy, and bigger muscles that do more work. This generates more heat, which needs to be piped out from inside of a more massive body. So

bigger bodies are all about structure; we have to be big enough to support a large brain but small enough to take it where we're going. So the size that we are may be optimum for a sapient species on the surface of a planet with Earthlike gravity.

33. The most common inexpensive 3D rendering is the anaglyph, where you project the left image of a stereo pair as red and the right image as blue, and then put on corresponding filtered glasses that let in only red light on the left and blue light on the right. Your brain combines the two sets of data as a single 3D black-and-white object.

34. Using 3D seismic imaging (tomography and migration) applied to radar echoes, we can in principle do something quite similar to a medical CT scan or a high-definition ultrasound scan of a primitive body like a comet nucleus or small KBO. It would be an easy science mission. P. Sava and E. Asphaug, "3D Radar Wavefield Tomography of Comet Interiors," *Advances in Space Research* 61, 2018.

35. William Blake, "Auguries of Innocence" (c. 1803).

36. The discovery of the imperceptible fact depends on advancements in physics and engineering—for example, focusing an ion beam to nanometer accuracy or flying a gigantic telescope into space. The James Webb Space Telescope (JWST) is designed to self-assemble in the L_2 stability region beyond the Earth from the Sun, its segments joining to form a 6.5-meter mirror, over three times the diameter of the Hubble Space Telescope. Since there is no gravity or wind or atmosphere to worry about, and since temperature can be accurately regulated by a sunshade, the banes of gigantic optical mirrors can be taken care of—the trick is getting them up there. It may sound strange, but optical telescopes many kilometers across are possible, and perhaps even likely, using fleets of micro-spacecraft holding mirrors in a coordinated arrangement, to reveal features of the universe and its planets that are presently not even imagined. The scientific future rides along with the technology of perception.

37. The basic photosynthesis equation, driven by sunlight and producing glucose, is $6CO_2 + 6H_2O \rightleftharpoons C_6H_{12}O_6 + 6O_2$. The weathering of rocks also relies on CO_2, which when dissolved in H_2O (e.g., the ocean) produces a weak carbonic acid solution, H_2CO_3, which breaks down minerals into clays and carbonates. If there was no photosynthesis, then any free oxygen would get used up producing seafloor muck out of the rocks that are eroded into the sea by the rivers. So the presence of free oxygen in a planetary atmosphere is an indicator that photosynthesis is happening, although there are abiotic ways of producing free oxygen as well.

38. A mass spectrometer is one of the most brilliant inventions ever. The basic idea is that you ionize (take an electron away from) an atom,

giving it a unit of charge. Then you apply a magnet to a beam of these ions, and the beam will deflect. The more massive the atom, the less it deflects for a given charge. So you basically weigh the mass of the atom and find out that some of them have one or two more neutrons than others. Most have one or two stable isotopes—that is, isotopes that don't break down radioactively into other elements. A particularly interesting one is xenon, which has eight stable isotopes (^{132}Xe, ^{129}Xe, ^{131}Xe, ^{134}Xe, ^{130}Xe, ^{128}Xe, ^{124}Xe, ^{126}Xe). Because xenon is a noble gas and doesn't react with anything, the ratios of these atoms to one another remain constant throughout the geologic evolution of a planet and its atmosphere, except when there is a process that fractionates them by mass. Each atom and its isotopes has a different kind of story to tell: xenon's tale is one of atmospheres, oxygen's is one of oxides (rocks and water), hafnium's is one of core formation, and lead's is one of crystallization.

39. How oxygen became diverse is much debated. The Sun's oxygen is 7 percent lighter (enriched in ^{16}O) relative to any of the meteorites and rocks from Earth, the Moon, and Mars. So some process in the nebula removed the lighter oxygen by a significant amount, perhaps analogous to how the freezing and thawing of ice sheets on Earth can remove and add lighter oxygen. This removal process would have happened differently here than there, producing different reservoirs of oxygen at Mars distance as opposed to Earth distance—at least that's the idea.

40. And for completeness let's not skip the possible step of a large-enough fragment to become a Mars-derived NEO, which then gets disrupted to deliver Martian meteorites to Earth.

41. What about meteorites ejected from the Earth, reimpacting the Earth in the form of meteorites, thousands or even millions of years later? Rocks are easily ejected from the surface of the Moon in modest-scale cratering events, and are likely to hit the Earth if they escape lunar gravity. Earth is different. It has a massive atmosphere, so only an improbably large impact crater will eject massive amounts of debris that will escape its substantial gravity. Nearly all of this would impact the Earth or be scattered in a million years, so unless a big crater formed on Earth quite recently (this has not happened), there will be few if any meteorites from Earth. As will be discussed later, the ejecta from ancient large cratering events on Earth would more likely be preserved on the geologically inactive Moon.

42. Yes, it's a word. It's what happens when geologists can't decide if they're biologists, but spend most of their time analyzing samples in a chemistry lab.

43. What about Mars? you ask. Well, Mars is so cold that it's not the triple point of water that governs its climate, but the freezing point of carbon

dioxide. That is what its atmosphere is made of, and some of it solidifies each winter, freezing out as a polar layer. Water on Mars is almost entirely in the solid phase, and there's not much of it.

44. Latin *wisdom,* although *Homo sapiens'* mileage may vary.

45. The LROC image browser and home page at Arizona State University is here: http://lroc.sese.asu.edu.

46. The HiRISE image browser at the University of Arizona is here: https://www.uahirise.org.

47. Pixel scale is how many meters across is one pixel of an image. This is not the same as image resolution, which is about two or three times the pixel scale, defined as the smallest features in an image you can distinguish.

48. NASA requires that all mission science data be archived and made permanently available to the public, usually within six months of the end of mission.

49. About equal to the amount of video data that was uploaded to YouTube per day in 2016.

50. A thousand years is actually about thirty "greats."

51. As recounted by the mission Principal Investigator Alan Stern and co-author David Grinspoon in *Chasing New Horizons: Inside the Epic First Mission to Pluto* (Picador: New York, 2018).

CHAPTER 1: RUINED STRUCTURES

1. Because science has been almost exclusively the domain of men, these signposts are almost exclusively men, until the 1900s. The marginalization of ambitious and creative female scholars limited what science could be and what it could achieve. This is changing, and its impact will be historical.

2. Nicolaus Copernicus avoided repercussions that would have been severe by publishing it shortly before his death. *De revolutionibus orbium coelestium* (*On the Revolutions of the Heavenly Spheres*), published in Nuremberg, Holy Roman Empire, 1543.

3. Then, as now, data ruled supreme: When Tycho Brahe died in 1601, Kepler inherited all of his previous master's observations, which was the edifice upon which he founded his theory of planetary motion.

4. Thirty years later, when Galileo was prosecuted for his proclamations that the Earth goes around the Sun, he recanted and got to experience thousands more orbits. His works were banned and he had to live under house arrest until his death in 1642. "And yet, in fact, they move," he is said to have maintained in private conversations.

5. Here's an excerpt from *Somnium,* translated by Normand Falardeau (MA thesis, Creighton University, 1962), where Kepler describes those

who can make the voyages to the Moon (Levania). One gets a taste for why Jonathan Swift was so fascinated by Kepler, and one sees some of the inspiration for Swift's *Gulliver's Travels*.

> The island of Levania is located fifty thousand German miles high up in the air. The journey to and from this island from our Earth is very seldom open; but when it is accessible, it is easy for our people. However, the transportation of men, joined as it is to the greatest danger of life, is most difficult. We do not admit sedentary, corpulent or fastidious men into this retinue. We choose rather those who spend their time persistently riding swift horses or who frequently sail to the Indies, accustomed to subsist on twice-baked bread, garlic, dried fish, and other unsavory dishes.

That would be the astronauts. And, showing that he understood acceleration and inertia, here's what they experience as the journey begins:

> The initial shock is the worst part of it for him, for he is spun upward as if by an explosion of gunpowder and he flies above mountains and seas. On that account he must be drugged with narcotics and opiates prior to his flight. His limbs must be carefully protected so that they are not torn from him, body from legs, head from body and so that the recoil may not spread over into every member of his body. Then he will face new difficulties: intense cold and impaired respiration. These circumstances which are natural to spirits are applied by force to man. We go on our way placing moistened sponges to our nostrils.

6. Only five solid shapes can be made out of N identical regular polygons, where N can be any number. First is the tetrahedron, 4 triangles, 3 to a corner. The cube is 6 squares, 3 to a corner. The octahedron is 8 triangles, 4 to a corner; the dodecahedron is 12 pentagons, 3 to a corner; and the icosahedron is 20 triangles, five to a corner. That's it; no more.

7. There are hints of the existence of Uranus in astronomical records from ancient Greece to China, as it is visible (sometimes, and barely) to the naked eye. But Kepler seems to have been unaware of it.

8. Houses built above granite need to be checked for radon gas, which is one of the elements on the decay chain from uranium to lead. Also, in case you are wondering, Pb stands for the Latin *plumbum*, hence *plumber* who works with lead piping and solder.

9. The force of gravity between two bodies of masses m and M, separated by distance r, is $F = GmM/r^2$. Here G is the universal gravitational constant, a number that seems to be constant throughout the Universe.

The force holding you to your chair is your mass m, times the mass of the Earth M, divided by the radius of the Earth r squared, times the constant $G = 6.6741 \times 10^{-11}$ m^3 kg^{-1} s^{-2}. The units, m^3 kg^{-1} s^{-2}, are what's needed to give you a force when you multiply it times mM/r^2, where the masses are in kilograms (kg) and the radius of the Earth is in meters (m). If you weigh 100 kg, the force in your chair is 981 kg m/s^2, which is abbreviated as 981 newtons. Divide that force by your mass, and every kilogram of your body feels a force of 9.81 newtons. Newton's second law states that force = mass × acceleration, $F = ma$, so in other words, the acceleration you feel on the surface of the Earth is $a = F/m$, or 9.81 m/s^2. So after one second of free fall, an apple is going to hit you on the head at a speed of 9.81 meters per second.

10. The perihelion of Mercury (when it is closest to the Sun) is $a_p = 0.31$ AU, while its aphelion (farthest distance) is $a_a = 0.47$ AU, so the *eccentricity* of the orbit is $e = (a_a - a_p)/(a_a + a_p) = 0.21$, where a perfect circular orbit has $e = 0$, and Earth has $e = 0.017$. You can make a perfect ellipse using two pushpins and some cardboard, and a piece of paper to trace on. Put one pushpin through the paper where the Sun shall be, and another pushpin some distance away representing the eccentricity. Put a loop of thread around the pins with some slack, and hold the thread taut with a pencil, and you will end up drawing an ellipse.

11. Newton was a devout Christian all his life, proving that you can break your worldview without shattering your faith.

12. Kepler's third law of motion, in terms of Newton's law of gravitation, is $P^2 = 4\pi^2 a^3/G(M+m)$, where P is the period (how long one orbit takes), a is the radius of the orbit, G is the gravitational constant $6.67 \cdot 10^{-8}$ cm^{3}g^{-1}s^{-2} and M and m are the masses, where m can be ignored if M is the dominant body. The more massive the planet (the bigger the M), the shorter the period (faster orbit), the way that a rock in a sling spins faster if you pull harder on the string.

13. Philip C. England, Peter Molnar, and Frank M. Richter, "Kelvin, Perry, and the Age of the Earth." *American Scientist* 95 (2007): 342–49.

14. Kelvin was a devout Christian who sought to adhere to the literal precepts of the Bible while taking physics to its logical conclusions.

15. The radioactive elements get concentrated in crustal rocks because when Earth's mantle solidified, the radioactive atoms were *incompatible*— that is to say, they find no space in the kinds of crystals that are forming. (A more familiar example is salt water: when you freeze it, it gets saltier because there is no room for salt in the ice crystals.) So they mostly remained in the liquid. The eruptions and intrusions that created the crustal rocks are therefore enhanced by these incompatible radioactive elements, much the way the lunar magma is concentrated in uranium and thorium (KREEP) as discussed below.

16. He estimated the tidal dissipation inside the Earth based on its current configuration with continents and oceans. There is a lot of drag pulling those oceans onto and off of the shore, so there is strong dissipation. So the rate of orbital expansion is high. But for most of the geologic history of the Earth, it has had only one supercontinent and one superocean, or just an ocean without continents, so dissipation has been lower.

17. Lead contamination was a problem with these measurements, leading to a large variation in calculated ages, so Patterson developed a clean-room protocol for studying lead. As a result, he became a key leader in demonstrating the widespread nature of industrial lead in the environment, which had not been known until then, leading to the elimination of toxic lead in gasoline, paint, and municipal water systems.

18. Zircon is a silicate mineral made of the compound zirconium, $ZrSiO_4$, that forms at high temperature and is preserved in some of the oldest rocks, not only on Earth but in lunar samples and ancient meteorites. When a zircon crystallizes, uranium is incorporated at measurable levels, but lead is not, making it an excellent chronometer. (Uranium is *compatible* with the crystal structure, substituting for Zr and thus making it into the crystal. Lead is *incompatible* and does not find a home; any lead you find in the crystal is therefore radiogenic.)

19. Another approach is called fission-track dating, where the uranium atoms inside of silicate crystals break down and emit decay particles that damage the local crystal lattice. Fission tracks can be counted under a microscope as a record of time passed, for billions of years. You can also look for fission tracks caused by cosmic rays; these can tell you how long a meteorite has been sitting on the surface of an asteroid or the Moon or another airless body, exposed to fast neutrons and other energetic particles that penetrate tens of centimeters of rock and cause radiation damage. Which is, incidentally, why astronauts need the shielding equivalent to about a meter of rock inside any long-term home.

20. It's funny, though, that we are able to know the sequence of events that happened in a small valley, for a couple of hours, on the day that the K/T impactor struck!

21. It is named after the man who popularized the theory—Johann Bode, director of the Berlin Observatory. It was originally published by the natural philosopher Johann Titius of Wittenberg University. It is often called the Titius-Bode law.

22. When the planets are not resolved—that is, when they are just points of light—you can tell how big they are by how bright they are, because the more area they have, the more light they reflect from the Sun. But that depends on their *albedo,* how bright or dark their materials (e.g., snow or dirt or charcoal) are. Ceres is considerably darker than the Moon, and only a fourth its diameter.

23. We haven't seen most of the planets in many of the systems. Because the data are incomplete, with enough free parameters (e.g., the way a leading 0 is tagged arbitrarily to the progression 3, 6, 12, . . . to account for Mercury in Bode's law) and allowing big enough measurement or other errors, and with allowances for special exceptions (e.g., asteroids) or lack of data, you can make a Bode-type law for almost any system. For an exoplanetary system detected by astronomical transits, we might not observe a Neptune-sized planet that has an inclined orbit to the rest of the system. One research paper has made predictions, using Bode's law (or a variant) to predict where to look in the data to find missing planets, and in several cases, they have been discovered that way!

24. This question keeps me up at night: Is there such a thing as "now" when we're always in the past from someone else's perspective?

25. *Galaxias* was the Greek word for the Milky Way, derived from the Greek *gala,* meaning milk, after Hera's milk that was spilled in the sky.

26. Basically, every gas cloud has a speed of sound—for instance, the speed of sound in air on Earth. This speed is a function of the temperature and density of the gas. If the gas is "stiff," with a speed of sound faster than the collapse, the wrinkles will smooth themselves out faster than clumping occurs. Setting these equal to each other (gravitational collapse equals the speed of sound) gives a characteristic size scale that governs the masses of the blobs that become the stars. That's how a lot of astrophysics and planetary geophysics works, comparing this to that, to get a characteristic size scale or time scale.

27. A cosmic year, one galactic rotation, is about 225 million to 250 million years. So in 4.6 billion years, we've gone around about eighteen to twenty times.

28. The Magellanic Clouds are our largest satellite galaxies. They were unknown to early European astronomers, and until the advent of astrophotography, to see them you would have to travel far, with expensive equipment.

29. The closest independent galaxy, Andromeda is three times more massive than our own, with about 1 trillion stars. Although the Universe is expanding, Andromeda is randomly moving toward us, at a speed of about 110 kilometers per second. That means it will collide with our galaxy in 4 billion years, sometime before the Sun enters its red giant phase.

30. A more mathematically detailed variation on this model, including a prediction for planetary masses, was worked out by the famed French mathematician and physicist Pierre-Simon, marquis de Laplace in the 1790s.

31. Ionized: stripped of one or more electrons so that the atom acquires a net positive charge.

32. Hal Levison (co-inventor of the Nice model and principal investigator of the Lucy mission) once reminded me, "Give a man a hammer, and everything looks like a nail." To which I replied, "Everything *is* a nail!"

33. As discussed below, there is not a continuous spectrum of light from a star, but absorption and emission lines corresponding to specific transitions of energy in the atoms of helium, hydrogen, calcium, iron, and other elements. So to quantify a cosmic redshift, one matches up recognizable spectral lines.

34. According to modern cosmology, space is increasing with time everywhere. The light is not emitted at a longer wavelength relative to us, as would be the case of a siren coming down the street, its tone shifting up as it approaches and shifting down as it leaves. Instead, the waves are spreading out as they travel because the space through which they are traveling expands. The longer they travel through space, the longer they get stretched out. So it's called the cosmic redshift, different from a Doppler shift.

35. Ironically, Hubble's first estimate for this constant was an expansion almost ten times as fast, implying a Universe less than 2 billion years old. That was problematic because by the late 1920s, radiogenic lead dating was proving that the oldest Earth rocks and meteorites were twice as old. And from another branch of astrophysics, 2 billion years proved much too short to account for the fact that the oldest stars (based on thermonuclear evolution models) seemed to be tens of billions of years old! It turns out Hubble had misinterpreted the distances to the galaxies by a factor of 10 or so, not recognizing that some of the objects in the neighboring galaxies that he plotted as stars were actually clusters of thousands of stars.

36. The radioisotopic chronology of meteorites and the dynamical modeling of accretion described later both give consistent answers of about 1 million to 3 million years (Myr) for the first planetary cores like Mars and Mercury, and then 10 to 100 Myr for the giant impacts to create the Earth and Venus. The most common early meteorites, *chondrites,* formed in the first 0.3 to 3 Myr, and the very oldest meteorites, which include many of the irons and the calcium-aluminum-rich inclusions (CAIs) inside the most primitive chondrites, date to less than 0.3 Myr.

37. Of course, if we do not *see* them (that is, resolve them) then we have to interpret what the infrared bump really means. The simplest answer is a protoplanetary disk. But "alien megastructures" have even been proposed for some of the more irregular observations. I'm skeptical of that. Star formation, just like planet formation, creates rings of material, and the material would be heated in a manner that well matches the thermal infrared observations. The weird, time-varying systems—well, they are not alien megastructures in stages of construction (maybe one

or two in the Universe might be that . . .) but just another system in the process of bashing its young planets together.

38. The telescopes (7- and 12-meter dishes) are moved around into optimal arrays depending on the nature of the observation—for example, whether you want better detection or better image resolution. Each telescope is actually a precise gigahertz radio antenna, whose detections are phased into images.

39. The spectral lines of iodine and other gases are shifted, in the starlight, relative to the spectral lines in a reference gas on Earth.

40. Discovered by Swiss astronomers Michel Mayor and Didier Queloz in 1995, reported in "A Jupiter-Mass Companion to a Solar-Type Star," *Nature* 378, no. 6555 (1995): 355–59.

41. One area where amateur astronomers have made a substantial difference is in follow-on observations of predicted transits. Small variations in transit timing can be measured using a small telescope, and are caused by the gravitational forcing of other planets in the system.

42. Resolution of 50 billionths of an arc second would be required to detect 100-kilometer features on a planet around a star that is 30 light-years away. Telescope resolution is the inverse of the telescope diameter, so the array of telescopes would have to be at least 20 kilometers across, and they would have to be located to sub-wavelength precision, less than one ten thousandth of a millimeter. Moreover, because the planet would be a billionth as bright as the star it orbits, each telescope in the array would require an occulting disk or some other technique to block out the starlight. That might seem impossible, but what today is routine astronomy was a near miracle thirty years ago.

43. Or a binary or triplet star. Binary stars are actually somewhat more common than singleton stars like the Sun; we see only a few of them in the sky because for the most part, one of the stars is many times dimmer than the other, or they are close together.

44. There appears to be a tendency for higher-metallicity stars to have gas giant planets orbiting close in. This might be because solids condense early when there's lots of material available, and these solids become seeds that accrete the hydrogen and helium before it can be blown away by the stellar wind.

45. If a star's exterior photosphere is well mixed, then its metallicity is representative of its composition as a whole, apart from the fusion products in its core. This in turn may be representative of the composition of the molecular cloud that made it and of the system of planets that formed around it.

46. According to astronomer Jonathan Fortney, about 1 percent of sunlike stars may have C:O of around 0.8–1.0, in the journal article "On the Carbon-to-Oxygen Ratio Measurement in Nearby Sunlike Stars:

Implications for Planet Formation and the Determination of Stellar Abundances," *Astrophysical Journal Letters* 747 (2012).

47. In the case of a carbon planet, the upper crust might be weaker than the lower crust, graphite over diamond, perhaps analogous to the situation of an ice sheet on Earth, on top of a granitic shield.

48. Ironically, it is our *artificial* net that is most vulnerable today to the effects of a coronal mass ejection (CME), because the electromagnetic pulse associated with it can shut down vast sections of the power grid for weeks to one to two years. In 1859, the largest CME in modern history caused sparks to fly at the telegraph offices and created spectacular auroras. In 2013 the insurance firm Lloyd's of London estimated the cost of a similar CME in the United States would be $0.6 to $2.6 trillion.

49. Pierre Kervella et al., "The Close Circumstellar Environment of Betelgeuse V. Rotation Velocity and Molecular Envelope Properties from ALMA," *Astronomy & Astrophysics* 609 (2018).

50. The superscript refers to the number of protons plus neutrons in the nucleus, known as the *atomic mass*.

51. Kilonovas might be responsible for the production of most of the heavier elements such as gold and molybdenum throughout the cosmos.

52. The first gravity wave was detected by the Laser Interferometer Gravitational-Wave Observatory (LIGO, NSF/Caltech/MIT) in September 2015, from a kilonova explosion caused by two neutron stars merging more than 1.3 billion light-years away.

CHAPTER 2: ROCKS IN A STREAM

1. Aristotle, *Posterior Analytics.*

2. Like Confucius and other classical figures of deep antiquity, Pythagoras is remembered somewhat as a man and somewhat as representing a period or school of thought. One of the legends is that Pythagoras was a mortal son of Apollo.

3. We say that towns like Syene are on the Tropic of Cancer, around 23 degrees north of the equator, or the corresponding Tropic of Capricorn in the south, where the Sun reaches the zenith at noon on the solstice.

4. In its heyday in the 3rd and 2nd centuries BC, the Library of Alexandria housed tens or hundreds of thousands of Egyptian writings in mathematics and the natural sciences, and thousands of translated works from Babylonia and Africa. Chinese and Indian scholarship going back thousands of years is similarly fragmented. The Babylonians, especially by the 8th and 7th centuries BC, were the most advanced in their time and made the most detailed astronomical records, for instance discov-

ering a 223-month frequency to the lunar eclipse. They introduced the measurement basis to astronomy, and timekeeping, and angles. (When Alexander the Great captured Babylon in 331 BC, the astronomer Kidinnu was put to the sword, apparently for resisting the translation of their astronomical tables.) In the West, there survive only hints of the profound astronomical legacies of the ancient Mayans, the Aztecs, and the Incas; these and other American cultures were obliterated by the Spanish, and nearly all the written records were intentionally destroyed; all that survives in many cases are the mounds. Likewise, in Cambodia and Central Africa, we can only guess what they thought about astronomy and mathematics from hints in their ruined temples and carvings and tablets.

5. Eratosthenes's estimate was based on the time it took walkers to go a certain distance, and thus was prone to error. Direct measurements were possible using ropes, but only for short distances. The analogous measure to the Greek *stadion* is the English *furlong,* which is the length of a plowed furrow, an eighth of a mile, which is ten *chains.* (One chain times one furlong is one acre. As its name implies, the chain was measured with a set of 100 *links,* which surveyors would carry around.)

6. The natural unit of angle is the *radian,* which is defined as the distance of arc along a circle, divided by the radius of the circle. So 2π radians is one full circle, 360 degrees. The size of the Moon divided by the distance of the Moon equals the angular diameter of the Moon, measured in radians. The Moon is half a degree across, so this is half of $1/360$ of 2π, about $1/110$ of a radian. Thus the Moon's diameter is $1/110$ of its distance.

7. If the Earth is a basketball under the hoop on a standard-sized court, then to scale, the Moon is a tennis ball on the three-point line. The Sun is a public swimming pool a mile to the east. Major asteroids in near-Earth space are grains of salt sprinkled on the court. Asteroids in the Main Belt are a few miles away, some of them as large as grapes or raisins, but most of them sprinkled around like rice.

8. The owner of a local restaurant I frequent is, according to the waiter, an adamant believer that the Earth is flat. Also that the world is surrounded by a transparent shell, and that rainbows are caused by sunlight refracting through the shell. Maybe he believes the Sun is pulled by Helios. (Yes, it is a pizzeria.)

9. The short letter summarizes a much longer treatise of findings that itself has been lost.

10. In addition to inventing exponentiation, Archimedes invented how to do math with numbers of arbitrary size. The division of two huge numbers is obtained by simply subtracting their powers: $10^a/10^b = 10^{a-b}$, and

their product is obtained by adding their powers, $10^a \cdot 10^b = 10^{a+b}$. (Note that there is a discrepancy between the value Archimedes arrived at and the number I got by using his estimated size of the Universe and the diameter of a sand grain, but that's a moot point, since his manuscript is lost.)

11. The first, you can prove by writing the equation $1/2 + 1/4 + 1/8 + \ldots = x$, then doubling it: $1 + 1/2 + 1/4 + \ldots = 2x$, and subtracting the equations. Voilà, all terms cancel and $x = 1$. For the second, you start with a square of paper and make four equal-sized squares, and now, divide those into equal-sized squares, and so on, and see if you can prove it!

12. When solar activity is high, energetic solar protons are a danger to astronauts, but they are a boon for exploration because they probe what's in the upper meter of the regolith that they can penetrate. Where there is water of any kind present in the upper meter (e.g., in shadowed polar regions or high latitudes), then there will be a diminished reflection of the solar protons, because an incoming proton is more likely to collide with another proton—namely, a hydrogen atom in water. That would damp its velocity, keeping it from bouncing back. Hitting a massive atom would reflect it. So it is from solar protons impacting the lunar surface, many of them implanted as hydrogen, that we know the Moon has, in places, a lot of subsurface water.

13. If you are unfamiliar with *Micrographia*, I suggest you go read it, and meet me back here: http://www.gutenberg.org/ebooks/15491.

14. The word *crater* comes from the Greek *krater*, a ceremonial wine cup. You have to provide the context yourself, whether a geologic feature is a volcanic crater or an impact crater.

15. Let D be the footprint of the camera image—that is, how many meters of terrain you see. The area of the footprint is D^2. Now, if the number of craters of size D per unit area on the planet goes as D^{-2}, then the number of craters per footprint is constant, no matter what your distance from the planet. The powers cancel, so this idealized landscape would appear similar at all scales.

16. We now know from computer simulations that the iron projectile would be scattered by the blast and little would stick around in the crater.

17. I have a couple of early cartographic maps of various regions on the Moon left over from an Apollo science study meeting that was held at the new UC Santa Cruz campus in 1967. On the maps are various pencil annotations that were made at the time—for example, that the chains of craters radiating from 93-kilometer diameter Copernicus are explosive "maar volcanoes." Today we know that these chains of craters are secondary impacts, caused by ejecta thrown out from the primary crater.

18. The first and only geologist to walk on the Moon was Dr. Harrison Schmitt of Apollo 17.

19. Survival of the weakest was promoted throughout the 1970s and 1980s by Donald R. Davis, the Tucson astrogeophysicist who among other things co-invented the idea of Moon formation by a giant impact. (He is not to be confused with Don Davis the space artist, whose work is featured above.) In responding to the idea of mega-craters on asteroid Mathilde, Davis invoked saguaro ribs embedded into the adobe walls of the forts of southeastern Arizona, as an analogy of how a porous asteroid might shelter itself against otherwise-calamitous collisions. See E. Asphaug, "Survival of the Weakest," *Nature* 402 (November 11, 1999): 127–28.

20. Margarita Marinova, Oded Aharonson, and Erik Asphaug, "Geophysical Consequences of Planetary-Scale Impacts into a Mars-Like Planet," *Icarus* 211, no. 2 (2011): 960–85.

21. Gregor Golabek et al., "Coupling SPH and Thermochemical Models of Planets: Methodology and Example of a Mars-Sized Body," *Icarus* 301 (2018): 235–46.

22. Synthetic aperture radar (SAR) creates a wave field by using the propagation direction of the spacecraft to construct an "aperture" from which to bring reflection images into focus. Yes, it's complicated, and it works.

23. He was ridiculed for his ideas when he proposed them.

24. The cycle of ocean basins opening and closing and continents reforming is called the Wilson cycle after Canadian geophysicist John Wilson. The time scale is around 200 billion to 300 billion years, several times the average age of oceanic crust.

25. This is all a mix of theory and conjecture, but is also supported by the data of *seismology*. Seismic velocity images show warm (low-velocity) material rising under rifts pushing them up, and dense (high-velocity) plumes sinking with the slabs. Seismology uses earthquake data to build an image of how fast elastic energy travels inside different layers and regions of the Earth. Waves propagate faster through cold solid slabs and in the solid inner core than they do through a partially melted mantle or through the liquid outer core. By processing a huge quantity of data from a lot of stations, the result is a 3D model of seismic velocity. So you can "see" (albeit very poorly) the low-velocity zones full of hot, mushy crystals, and the fast zones revealing cold, sinking slabs. The "slab graveyard" from early plate tectonics is inferred from a low-velocity seismic reflection zone at the core-mantle boundary.

26. One creative approach to Hadean geology is M. Santosh, T. Arai, and S. Maruyama, "Hadean Earth and Primordial Continents: The Cradle of Prebiotic Life," *Geoscience Frontiers* 8, no. 2 (2017): 309–27.

27. Chicxulub formed while the Deccan Traps were also forming in what is now central India, a large igneous province that would have also changed the atmosphere in equally profound but more gradual ways, releasing huge volumes of gases from the mantle. While it does appear that certain ecosystems were already on the decline, the K/T extinction was caused by the asteroid impact.

28. The impact was especially significant biologically because the asteroid struck into sediments that released considerable amounts of sulfate aerosols and small particulates into the atmosphere. The reaction of sulfate ions with water produces sulfuric acid, which reduced the oceanic pH to the point that calcareous planktons in the upper hundred meters were dissolved.

29. Massive limestone sediments were laid down on top of the Chicxulub crater following the impact, where today are found some of the most amazing caverns ever explored. The famous cenotes of Mexico, deep-walled limestone lakes that grace Mayan calendars in Tucson taquerias, are the result of karsts opening up in Tertiary limestone formations circumscribing Chicxulub. As the upper layers of the formation are eroded away, the kilometers-thick sediments expand, and this causes cracks and interconnected caverns. I suspect that the subterranean ocean of tidally worked and brutally cratered Ganymede has cavernous webs extending for thousands of kilometers.

30. Ultimately it was discovered in proprietary PEMEX seismology and gravity data (Petróleos Mexicanos, the Mexican state oil company). The detective story itself is remarkable, from the original discovery in the late 1970s by petroleum geologist Glen Penfield to the "discovery of the discovery" by (at the time) planetary science graduate student Alan Hildebrand.

CHAPTER 3: SYSTEMS INSIDE SYSTEMS

1. Hydrogen atoms, fusing into helium atoms, lose a little mass each time because the nuclear structure gets more tightly bound; this gives off energy according to Einstein's equation $E = mc^2$. By hydrogen fusion, the Sun is losing mass at 4 million tons per second, and generating energy. Our galaxy is losing about one solar mass as mc^2 every thousand years.

2. The time scale of impact cratering gets longer the larger the event—that's easy to understand. Other aspects of physics also "scale." The theory (or procedure) known as *dimensional analysis* is based on the fact that whatever equation you come up with, the units have to work out. If you want an equation for the velocity of something, the answer has to be in meters per second. Everything breaks down into funda-

mental units or dimensions: time (seconds, s), distance (meters, m), and mass (kilograms, kg). Other units derive from these: velocity (distance per unit time, m/s), density (mass per unit volume, kg/m³), gravity (velocity per unit time, m/s²), force (kg·m/s²), and so on. Suppose you want an equation for crater diameter (m) and all you know is the projectile velocity (15,000 m/s is typical) and the target gravity (1.6 m/s² for the Moon) and density (2,700 kg/m³ for lunar crust). The fact that the units have to work out forces you into a very narrow set of mathematical equations. Then you do experiments in the lab and fit the equations to the data, and the results are *scaling laws*. That's how scientists use laboratory experiments and data from nuclear weapons tests to predict the physics of impact cratering at scales of hundreds to thousands of kilometers.

3. This is how a reaction wheel works to control a spacecraft: a motor spins up a flywheel, so the spacecraft spins the other way, so you can point here and there and keep your solar panels facing the Sun.

4. With a little geometry you can show that the probability goes as $\sin(2\theta)$, where θ is the impact angle. This function peaks at 45 degrees, and goes to zero probability for perfectly head-on collisions, $\sin(0) = 0$, and for completely grazing collisions, $\sin(2 \cdot 90°) = 0$.

5. Newton's constant $G = 6.672 \cdot 10^{-8}$ cubic centimeters per gram, per second squared. It appears in the force law $F = GMm/r^2$ where F is the force of gravity between two bodies, one of mass M and the other of mass m, separated by a distance r. It seems to me there must be some reason that it appears to be that simple, or very closely so.

6. Swiss astrophysicist Alexandre Emsenhuber has shown that collision chains are more probable than single giant impact mergers, and that oligarchs can hop from target to target (e.g., from a hit-and-run with Venus, to then collide with Earth) a good fraction of the time. However the collision chain plays out, it is the final merger that ends up forming the protolunar disk, because the disk gets spun out primarily by the angular momentum of the merging cores. This means that the standard model of Moon formation is probably right, since in order to do accretion the collision *had* to be slow, and was most probably around 30 to 60 degrees. But you have to factor in the previous giant impacts between proto-Theia and the proto-Earth that could have mixed the two bodies already; there might have been one or even two of them. In fact maybe Theia came from Venus after a hit-and-run collision there.

7. Water chemistry is represented by factors such as pH and salinity. Most of Earth's surface is an open salty ocean, slightly basic, pH = 8.1 and dropping, where 7 is neutral.

8. Some labs study the physics of ices at low-pressure conditions, and other labs study the physical and thermodynamic properties of water

and ices and brines, and other materials, under high-pressure conditions, using a high-pressure anvil cell to control temperature and pressure to attain planetary interior conditions. Gem-quality diamonds are used for the anvils, and then a laser beam can be transmitted through them to probe what is going on.

9. In addition to the phases of pure ice, there are *clathrates,* in which water molecules act as solid cages holding methane or other molecules inside. Methane clathrate is stable at depth in the cold oceans, and if you bring up a chunk, it looks like ice but you can light it on fire!

10. Kurt Vonnegut, *Cat's Cradle* (New York: Delacorte Press/Seymour Lawrence, 1963). Ice-nine is a phase of water that when it comes in contact with any other water below a temperature of 45.8°C (this is fiction, mind you) serves as a catalyst that turns that water into ice-nine. It was invented by the military so that soldiers wouldn't have to deal with mud. I won't tell you how the book ends, but one of the final chapters is "The Grand Ah-Whoom."

11. Crystallization of water liberates energy, keeping it just at the freezing point. Likewise the melting of ice consumes energy without changing temperature, making it hard to raise the temperature above zero. If you put a thermometer into a glass of water in the freezer, it will cool down to 0°C over some period of time, depending on how much water at what starting temperature and how cold the freezer. Then at 0°C the first ice crystals start to form. The water continues to radiate energy into the freezer, which is colder than it is, but its temperature does not drop. What does this mean, to lose thermal energy but not decrease in temperature? That is the concept of *entropy,* as represented by the phase transformation from water to ice, giving off thermal energy to solidify, without changing temperature. The water goes from higher (more disordered) to lower (crystallized) entropy.

12. You can't cool off when it's humid because your sweat can't evaporate: the air is already saturated and can't take any more molecules of water.

13. Some areas are famous for frost heave of boulders, like the northeastern United States and places like Norway and southern Chile and Argentina. It happens when water in the warm season gets pulled toward the boulders, which are the coldest objects. It gets pulled there because it is dry, since it is still frozen, and water likes to migrate to where it is dry (the principle of equilibration). Then it freezes there and makes a thicker mass. That pushes the rocks up. When it thaws, mud and tiny grains fill the spaces that are left by the melting ice. So it ratchets the rocks to the surface over hundreds of years. Puzzlingly, this rock-cycling process appears to happen on Mars, but it seems to require liquid water, or conditions close to it, which is way above the regolith temperature Mars has experienced for billions of years. The general word

for it is *cryoturbation*. When animals (e.g., gophers and ants) overturn the soil, it is *bioturbation*, which we don't think is happening at Mars!

14. After the fairy tale about the girl Goldilocks, who came to the house of the Three Bears and found one of their beds, one of their chairs, and one of their bowls of porridge to be *just right*. Astronomers have adopted the story to refer to the habitable zone.

15. If you wanted Mars and Venus to also be more Earthlike, you might just swap their places. Move Venus, sweltering under 90 bars of atmospheric CO_2, out twice as far, where it will get one-quarter of the incident radiation. Bring Mars from 1.5 AU to 0.7 AU and its water and CO_2 ice will melt, kick-starting a bigger atmosphere.

16. One bar is about equal to the atmospheric pressure on the surface of the Earth, 1 atm = 1.01 bar. In its metric definition, 1 bar = 100,000 Pa (pascal), where in turn 1 Pa is 1 newton of force applied on 1 square meter. In British units, the surface pressure of the Earth is 14.7 pounds per square inch (psi), which is the weight of a 760-millimeter column of mercury (Hg), which is how they used to measure it. So you often see pressure reported as mmHg.

17. The basic equation for photosynthesis is $6CO_2$ + $6H_2O$ + light energy = $C_6H_{12}O_6$ + $6O_2$, where the big molecule is glucose.

18. The O–C–O molecular structure of CO_2 has bending and stretching vibration modes that resonate with the frequencies of the thermal infrared photons emitted from a planet's surface. So it absorbs them, and the atmosphere heats up. The V-shaped H–O–H structure of the water molecule has even more vibrational modes in the infrared, thus is an even greater plague to astronomers, but they can avoid most of it by getting on top of the highest mountains.

19. The increasing CO_2 in the atmosphere acidifies the oceans, which is the primary threat to coral reefs throughout the world, and to calcareous planktons, as well as to mollusks.

20. Vinegar, a mild acid, dissolves it.

21. But not too deep—below about 4,000 meters' depth, the solubility of $CaCO_3$ abruptly increases. Deeper than this *carbonate compensation depth*, the carbonate remains dissolved in ocean water.

22. If Mars ever had a massive CO_2 atmosphere that disappeared—as has been proposed to explain a "warm, wet" ancient climate—then it too should have carbonates, which have not been found.

23. The basic facts have been known for more than a century. The first work on CO_2 and the greenhouse warming of the Earth was by the Swedish chemist Svante Arrhenius, "On the Influence of Carbonic Acid in the Air upon the Temperature of the Ground," *Philosophical Magazine and Journal of Science* series 5, 41 (1896): 237–76. The alarming effect of industrialized CO_2 on Earth's climate has been

securely known for over forty years. Why has nobody listened to the scientists?

24. A circular orbit has the lowest energy for a given orbital angular momentum.

25. The fundamental paper handed out to graduate students on this topic is "Q in the Solar System" by California geophysicists Peter Goldreich and Steven Soter.

26. Stephen Vance et al., "Ganymede's Internal Structure Including Thermodynamics of Magnesium Sulfate Oceans in Contact with Ice," *Planetary and Space Science* 96 (2014).

27. Large planets have a lot more accretional heating to begin with, and produce a lot more radioactive heat. But what's more important is that it takes longer for the heat to get out, which is why Jupiter is still cooling off while the Moon solidified in a few million years. A super Earth might take a billion years to lose its original heat, and would continually produce radioactive heat that would have to get out through its oceans. Planets are giant heat engines.

28. From the ESA website, http://sci.esa.int/juice/50074-scenario-operations/:

> The JUICE spacecraft will be launched in June 2022 by Ariane 5 and will use Venus and Earth gravity assists in its 7.6 years cruise to Jupiter. After the orbit insertion in January 2030 the spacecraft will perform a 2.5 year tour in the Jovian system. [. . .] During the tour, gravity assists with Callisto and Ganymede will shape the trajectory. Two targeted Europa flybys are included focusing on composition of the non water-ice material, and the first subsurface sounding of an icy moon. [. . .] The mission will culminate in a dedicated eight months orbital tour around Ganymede during which the spacecraft will perform detailed investigation of the moon and its environment and will eventually impact on Ganymede.

29. The formula for kinetic energy is $E = \frac{1}{2}mv^2$, where m is the mass of your capsule and v is the velocity. This has to get turned into heat, or else the velocity must be transferred into the target (what happens is a bit of both) in order for the capsule to stop. The higher the velocity, the more kinetic energy there is, until the result looks just like an explosion. Atmospheric entry is a controlled explosion in the atmosphere, achieved by surrounding the bottom (forward side) of the capsule in a shield of material that is designed to ablate away, shaped to take the heat energy away from the capsule. While the heat can be dissipated using an ablative shield of sufficient mass, the vibration is a different matter. Air is turbulent, and so just like an airplane coming in for a

hot afternoon landing, there will be pockets where deceleration is slow (low pressure) and pockets where deceleration is fast. This results in a change in acceleration per unit time (meters per second per second, per second, known as *jolt*), which discombobulates a living system. Much better to reserve this approach for hardened exploration probes than for humans!

30. A hydrogen or helium balloon won't work on Jupiter or Saturn the way blimps work on Earth, because the atmospheric gas is already hydrogen and helium. For a balloon to float, it has to be filled with a molecule lighter than the surrounding gas. A hot-air balloon would work; you'd use a tank of oxygen to combust with the available hydrogen.

31. We don't call it "liquid," because although it is dense and fluid, there is no distinct transition from the gas phase, just higher and higher densities.

32. The war in Iraq cost an estimated $3 trillion to $6 trillion. According to the U.S. inspector general, $9 billion was unaccounted for during one half year in 2003–04. So it's not like we can't afford it.

33. E. M. Shoemaker, R. J. Hackman, and R. E. Eggleton, "Interplanetary Correlation of Geologic Time," *Advances in the Astronautical Sciences* 8 (1963): 70–89.

34. Swapping among the other planets is far less common but also happens. To get to the Earth from Mercury and vice versa require a huge ejection velocity. Getting from Venus to Earth is easier but requires penetrating the massive atmosphere, something that last happened billions of years ago.

35. In one of the first detailed geophysics cruises to the area, German geophysicist Rainer Gersonde and meteoriticist Frank Kyte of UCLA discovered gravel-sized fragments of mesosiderite, a type of meteorite that has a large metallic fraction, embedded in sediments from drill cores pulled up from the muck 5 kilometers deep. (Kyte had earlier also found the "smoking gun" of meteorite fragments from the Chicxulub impactor.) It was an asteroid that did it, and a metal-rich one at that. Seismic arrays towed behind the ship showed widespread disturbances to the seafloor, consistent with a blast that excavates 4 to 5 kilometers deep, about a 1-kilometer asteroid traveling at 20 kilometers per second. Hundreds of millions of tons of ejecta, in this case seawater, would be ejected.

CHAPTER 4: STRANGE PLACES AND SMALL THINGS

1. This model is described by Erik Asphaug, Martin Jutzi, and Naor Movshovitz, "Chondrule Formation During Planetesimal Accretion," *Earth and Planetary Science Letters* 308, no. 3–4 (2011): 369–79.

2. Comets are named for their discoverers. Long names like Churyumov-Gerasimenko and Schwassmann-Wachmann are why comets often go by their numbers, in this case 67P and 29P.

3. Dante Lauretta et al., "The Unexpected Surface of Asteroid (101955) Bennu," *Nature* 568 (2019): 55–60.

4. According to astrophysicists Zoë Leinhardt, Robert Marcus, and Sarah Stewart, "The Formation of the Collisional Family Around the Dwarf Planet Haumea," *Astrophysical Journal* 714, no. 2 (2010): 1789–99.

5. It has not been taken kindly by the Pluto-is-a-planet people that there is a campaign to call the yet-to-be-discovered distant body, ten Earth masses and a thousand AU away, "Planet Nine." For this book I refer to any predicted planet as Planet X.

6. Just like Gauss obtained a most likely *orbit* of asteroid Ceres when he invented least squares, Eliot Young and his colleagues obtained the most likely *image maps* of Pluto and of Charon by populating the pixels in those blank maps with variations on the possible brightness and color values. They aligned their modeled planets at every observation time and adjusted the pixel values until they obtained the best fit to all of the telescopic observations of the whole planet (both bodies).

7. In a twist on this idea, Craig Agnor and Doug Hamilton showed in 1999 that Triton itself could actually be a captured KBO. Direct capture doesn't work, for many of the same reasons that you can't form the Moon by direct capture around the Earth. Their trick is to have the KBO be a binary pair, just like Pluto-Charon, that encounters Neptune. Once in a while a binary KBO would have come close enough to Neptune for it to be "ionized"—that is, the planet causes one body to go this way and the other body to go that way. And once in a long, long while, one of these ionized components will get captured by Neptune, while the more massive one escapes, perhaps later to suffer a collision with Neptune (explaining why it's gone). It may be most likely for ionization capture to happen when the encounter is retrograde.

8. A great theory can be both important and wrong.

9. An earlier name for the LHB was the late lunar cataclysm, a phrase that is a bit more objective because it does not presuppose it was a solar-system-wide bombardment.

10. See Barbara Cohen, Tim Swindle, and David Kring, "Support for the Lunar Cataclysm Hypothesis from Lunar Meteorite Impact Melt Ages," *Science* 290, no. 5497 (December 1, 2000): 1754–56. Despite the title, the paper's conclusion is that there is not a spike, but instead a broad distribution of impact melt ages in lunar meteorites ranging from around 2.7 billion to 4.2 billion years.

11. The original Nice model, that is. Like the giant impact theory, the Nice model is polymorphic.
12. From *Gulliver's Travels,* 200–201.

> The innermost is distant from the centre of the primary planet exactly three of the diameters, and the outermost five; the former revolves in the space of ten hours, and the latter in twenty-one and a half; so that the squares of their periodical times are very near in the same proportion with the cubes of their distance from the centre of Mars, which evidently shows them to be governed by the same law of gravitation, that influences the other heavenly bodies.

13. A modern version of this is to publish an idea in an obscure journal that nobody will read until the time comes to cite it. In the age of the internet, this has become impossible. I suppose a closer analogy might be an encrypted document, where you post your entire theory or discovery online and then later provide a key to unlock the result.
14. This is a classic counterexample to the epistemological question whether knowledge is "justified true belief." It is true that Mars has two moons. And Kepler believed it. And he was justified in believing it. But it wasn't knowledge.
15. Kevin Walsh et al., "A Low Mass for Mars from Jupiter's Early Gas-Driven Migration," *Nature* 475 (July 14, 2011): 206–9.
16. We used to joke in our planetary science journal club that the discussion of any paper wasn't finished until somebody had related it to the Nice model.
17. This tries to summarize a brilliant set of theories developed by Man Hoi Lee, Stan Peale, Robin Canup, Bill Ward, and several others; details are not agreed upon.
18. David Stevenson of Caltech has written extensively on the interior structures of the giant planets and why Jupiter is so different from Saturn.
19. Adam Showman and Renu Malhotra, "Tidal Evolution into the Laplace Resonance and the Resurfacing of Ganymede," *Icarus* 127, no. 1 (1997): 93–111.
20. Salt water is a good conductor of electricity because of all the ionized molecules, positively charged sodium (Na) and negatively charged chlorine (Cl). When Galileo measured the magnetic field of Jupiter, the field was perturbed around Europa, in the way that the field of a magnet is perturbed when you put it next to a conductor like a nail. Europa is conductive and it isn't made of metal, so the conductor is a liquid salty ocean.
21. Jupiter's magnetic field, looking at the planet as a bright dot from Earth,

extends as large as your fist held up to it in the sky. All its moons orbit deep within it.

22. There are even mission concepts to explore Europa using a nuclear thermal pack to melt a probe down through the ice, sinking and sinking and refreezing behind it, laying a communications tether. Another clever, bad idea; we should wait until we advance the technology to do it correctly.

23. This is, coincidentally, the age of a typical ocean basin on the Earth before it gets recycled by plate tectonics.

24. Graduate students remember them by "Met Dr. Thip."

25. One gigawatt is about half the power generated by Hoover Dam.

26. Titan orbits at 20 Saturn radii, while Io, Europa, Ganymede, and Callisto orbit at 6, 9, 15, and 26 Jupiter radii.

27. Erik Asphaug and Andreas Reufer, "Late Origin of the Saturn System," *Icarus* 223, no. 1 (2013): 544–65.

28. Matija Ćuk, Luke Dones, and David Nesvorný, "Dynamical Evidence for a Late Formation of Saturn's Moons," *Astrophysical Journal* 820, no. 2 (2016): 97.

29. The mission, led by planetary scientist Elizabeth Turtle, is described in *Dragonfly: A Rotorcraft Lander Concept for Scientific Exploration at Titan* (R. Lorenz et al., 2018, Johns Hopkins APL Technical Digest 34) and at dragonfly.jhuapl.edu. The craft is a quad-octocopter (4 dual rotors) powered by radioisotopic thermal generators (RTGs).

30. Space tourism to Mars is the elephant in the room at this point. Planetary Protection is an agreement between space agencies. If a private company wants to go to Mars, compliance with Planetary Protection standards is optional. There may be a narrow window to discover indigenous life on Mars before we bring our own.

31. An ion thruster uses electricity from solar panels, or from a small nuclear-powered generator, to ionize atoms of a heavy neutral gas like xenon (Xe), giving them a charge. Then the electricity is used to expel the ionized atoms to hundreds of kilometers per second, not producing a huge amount of thrust, but gradually giving the spacecraft a large acceleration (delta-v, change in velocity that gets you there). Chemical propulsion is fast; ion drive is efficient.

32. Titan Mare Explorer was proposed in 2009 by Ellen Stofan and her team to land in Ligeia Mare and drift in the currents and the winds. The mission was in final competition for selection but required nuclear power units that ended up not being available.

33. The Austrian physicist Erwin Shrödinger wrote a letter in 1935 to the journal *Naturwissenschaften* to express concern about the new "blurred model" (as he called it) for representing reality that was quantum me-

chanics. He imagined putting a cat in a steel box and closing the lid. "In a Geiger counter, there is a tiny bit of radioactive substance, so small, that perhaps in the course of the hour one of the atoms decays, but also, with equal probability, perhaps none; if it happens, the counter tube discharges and through a relay releases a hammer that shatters a small flask of hydrocyanic acid." After one hour, until we (the observer, whatever *that* means) open the lid, the cat exists equally in both states, alive and dead.

34. If you have a separate communications relay orbiter, then the boat can be more sleek and less of a windsail. It's probably cheaper and less complicated to include the antenna in the design of the boat; however, in gusty winds, this decision could be problematic.

CHAPTER 5: PEBBLES AND GIANT IMPACTS

1. You can plot all the current reduced exoplanet data in any way you want at http://www.exoplanets.org.

2. Asteroids are hottest not at the sub-solar point, but rotated a bit into the "afternoon," just like here on Earth. The higher-energy photons radiating from the hotter afternoon surface carry a teeny bit more quantum-mechanical momentum, so heat, ultimately due to the Sun, acts like a super-subtle thruster pointing in the afternoon direction. This can speed up or slow down the orbit, depending on the orientation of the spin pole. Over millions of years, this *Yarkovsky effect* can cause a small asteroid to drift by as much as a few AU, causing it to get caught into disruptive resonances with Jupiter or Saturn, transferring it from the Main Belt to the inner solar system to become an NEO. A related effect called YORP is driven by the same thermal physics, but is related to the fact that an asteroid shape always has some chirality (unless it is perfectly symmetrical, like a sphere). If thermal photons radiate from all over its surface, sort of like tiny jets, their thermal thrust will impart a net spin. Both of these very subtle forces turn out to be of critical importance to the long-term dynamical evolution of common-sized asteroids. The Yarkovsky effect was predicted by the Polish railway engineer Ivan Yarkovsky in 1901; solving scientific problems was his hobby. However, its importance to asteroids would not be recognized until the 1990s.

3. In case you are wondering, this thermal spin-up (YORP; see note above) can only spin up bodies smaller than a few tens of kilometers diameter over the age of the solar system. So it will not help with Darwin's problem of spinning up the Earth to lunar fission.

4. Hence, *meteorologist,* the person who presents the weather on TV.

5. British mineralogist Henry Sorby, "On the Structure and Origin of Meteorites," *Nature* 15 (1877): 495–98. He interpreted chondrules as "droplets of fiery rain" that he suspected came from the Sun.

6. William Thomson, Lord Kelvin, "On the Age of the Sun's Heat," *Macmillan's Magazine* 5 (1862): 388–93.

7. That is to say, the famous but probably made-up quote by Thomas Jefferson: "I would rather believe two Yankee professors would lie than that stones have fallen from the heavens."

8. Swift-Tuttle is a 30-kilometer beast discovered in 1862 whose orbit comes just inside the Earth's every 133 years (perihelion 0.95 AU) and then spends most of its time way beyond Neptune and Pluto (aphelion 50 AU). It has its next closest approach to Earth (only 1 million kilometers—0.01 AU) in 3044, and will someday probably collide with Earth. It is in a 1:11 resonance with Jupiter, which possibly makes it "exploitable," in a planetary engineering sense, for colonization of the outer solar system when we get to that point.

9. Fat-laden diet potato chips were briefly sold using right-handed fats, but the propensity for causing "anal leakage" curtailed the market for such foods.

10. A meter-thick gabion wall could be built out of asteroid regolith sieved to 10- to 30-centimeter sizes, the overall structure held down by lightweight webbing extending 1 meter from the living quarters. A 50-meter diameter asteroid (one-tenth the size of Bennu) could cover a 200-meter long by 100-meter diameter cylinder to a depth of 1 meter, so radiation shielding is not a problem.

11. The principle of equivalence is that you can't tell the difference between being in a small capsule accelerating at 0.001 g and being in the same capsule at rest on the surface of a small asteroid.

12. A 30-meter diameter meteorite, corresponding to the Tunguska explosion, would have left a detectable *iridium anomaly,* in the way that excess iridium at the K/T extinction boundary was proposed by physicist Luis Alvarez and colleagues to be the result of a 10-kilometer asteroid. The meteorite could have been icy (without iridium), or it could have been a differentiated mantle fragment (also without iridium), or too many years might have passed in the bitter-cold, mosquito-ridden swampy bog between the huge 1908 explosion and the first investigation in 1927.

13. Early in the 1970s, in his novel *Rendezvous with Rama*, Arthur C. Clarke envisioned a Spaceguard program in which robotic telescopes look out for any Tunguska-class events. The telescopes ended up discovering an interloping dormant alien spacecraft, in a plot that was echoed by the discovery of interstellar interloper 'Oumuamua by the Pan-STARRS telescope used for NEO surveys.

14. The average temperature inside a comet in the outer solar system is about 30K–40K (Kelvin). Absolute zero (0K) is the coldest temperature there is, when atoms stop vibrating, never actually attained, equal to minus 273°C. It is the physicist's zero point. Celsius is the water-based zero point (colder than 0°C it freezes, hotter than 100°C it boils). Fahrenheit is the human-based zero point (colder than 0°F or hotter than 100°F, you can't survive for long).

15. A triple system of massive distant bodies could be in a state of *forced libration* for many billions of years, maintaining liquid brine deep below with scarcely a photon hitting their surface.

16. The glowing tails of comets are based on similar physics as auroras—photons emitted by strongly ionized gas—but they do not show the same kind of movement because comets are much larger in extent and not subject to the chaos of Earth's magnetic field. The Ulysses space probe deduced that it must have flown through the tail of Hyakutake in 1986, leading to the astonishing result that the tail was 3.8 AU long—quite a structure for a 4-kilometer diameter nucleus!

17. A comet has an albedo of around 3 percent, which means that for every 100 photons (at the specified frequency) that are incoming, 97 of them are absorbed. At visible wavelengths, primitive bodies like comets and carbonaceous asteroids have albedos of around 3 to 5 percent, the color of charcoal. The albedo of the Earth is a complicated and changing (and important!) measurement but is around 30 percent. *Spectral slope* is associated with the albedo; an asteroid is *red* if the albedo is higher in the red than in the blue.

18. Kathryn Volk and Brett Gladman proposed in 2015 that Mercury was the ground-down relic of a set of massive planets interior to Venus, and Konstantin Batygin and Gregory Laughlin proposed in 2016 that there was a destroyed inner solar system. Both invoked Jupiter and Saturn moving in and laying waste to original planets, and then moving back out, the "grand tack."

19. William B. Yeats, "Easter, 1916."

CHAPTER 6: THE LAST ONES STANDING

1. As translated by Nathan Sivin, *Cosmos and Computation in Early Chinese Mathematical Astronomy* (Leiden, Neth.: E. J. Brill, 1969).

2. The Macduff hypothesis: "Despair thy charm, / And let the angel whom thou still hast served / Tell thee, Macduff was from his mother's womb / Untimely ripped."—Shakespeare, *Macbeth*.

3. Urbain Le Verrier at Paris Observatory and John Couch Adams at Cambridge University made the same calculations and arrived at fundamentally the same prediction, days apart, for the existence of

Neptune. Le Verrier was a few days faster out the gate, and when his colleagues at Berlin Observatory received his letter indicating its predicted location, they pointed their great refractor (telescope) and discovered the planet that same night.

4. The Large Synoptic Survey Telescope (*synoptic* = never closing its eyes) is being built on Cerro Pachón in Chile. With Pan-STARRS in the north and LSST in the south, the whole sky coverage will be complete at an image sensitivity and resolution that will easily detect a massive Planet X.

5. According to calculations by Robin Canup and me ("Origin of the Moon in a Giant Impact Near the End of the Earth's Formation," *Nature* 412 [2001]: 708–12) on the basis of the first giant impact simulations by Swiss astronomer Willy Benz.

6. In the proposed hit-and-run scenario, by Swiss astrophysicist Andreas Reufer and colleagues ("A Hit-and-Run Giant Impact Scenario," *Icarus* 221, no. 1 [2012]: 296–99), most of Theia keeps going after the giant impact. Did it come back to hit the Earth again, or is it inside Venus? Alexandre Emsenhuber of the University of Arizona and I are working on the latter half of the problem. If there was a "hit-and-run return" origin of the Moon, then the first giant impact would have given the Earth a lot of spin but kept going, and the second giant impact into Earth would be slower, hence accreting Theia, which spawns the disk that made the Moon. The double whammy mixes up the planets compositionally and makes a protolunar disk that is strongly inclined relative to the Earth's spin axis.

7. This is controversial too. Just because you can vaporize H_2O and other volatiles does not mean that you will lose them from the vapor cloud. The cloud would be gravitationally bound, so volatile loss is not a solved problem for the Moon.

8. Uwe Wiechert, Alex Halliday, and colleagues, "Oxygen Isotopes and the Moon-Forming Giant Impact," *Science* 294, no. 5541 (2001): 345–48.

9. Theories are reviewed in Asphaug, "Impact Origin of the Moon?" *Annual Review of Earth and Planetary Sciences* 42 (2014): 551–78.

10. He was writing on the broader topic of solar system formation, and the two competing hypotheses at the time plus their variants, one being the disk model, the other being a planet system pulled out of the mantle of the Sun by a passing star, the latter being at that time called the planetesimals hypothesis. Harold Jeffreys, "The Planetesimal Hypothesis," *Observatory* 52 (1929): 173–78.

11. This is the hypothesis of *synestia* advocated by planetary physicist Sarah Stewart of UC Davis. If high-impact velocities can be generated in a consistent manner, then powerful shock states can ensue, making

the explosion all the more intense. The result is a gravitationally unstable melt-vapor cloud that recondenses back to form a planet. My graduate student quipped, "If your synestia lasts longer than four hours, call your doctor."

12. That myth is actually a good analogy of the standard model of the giant impact, since Athena sprang from Zeus's forehead as a result of him swallowing her mother, Metis. Unfortunately we can't go renaming all the planets to match the best myths.

13. Named for French astrophysicist Édouard Roche, who developed the idea of a tidal breakup limit in 1848 and proposed that Saturn's rings were created when a small moon was catastrophically destroyed by tides.

14. The Martian day is 24.6 hours long and hasn't changed much, while the Earth's day has been slowing down from 5 hours at its formation, and now just happens to be about the same length as the Martian day. Its axial obliquity is also about the same as ours, so Mars has the same basic seasons as we do. (We write this all up to coincidence; should we?) For rover operations specialists, "Mars time" is sometimes very coincident with their local time, and after two months becomes exactly opposite, so if you want to be part of a team operating active missions on Mars, it helps to be flexible in your sleep schedule.

15. Making craters does not do a very good job at spinning up a planet, so Mars's rotation probably cannot have varied by too much since the formation of Borealis.

16. Since the problem requires higher numerical resolution, studies of the post-giant-impact disk around Mars are relatively few, including Margarita Marinova et al., "Geophysical Consequences of Planetary-Scale Impacts into a Mars-like Planet," *Icarus* 211, no. 2 (2011): 960–85; Robin Canup and Julien Salmon, "Origin of Phobos and Deimos by the Impact of a Vesta-to-Ceres Sized Body with Mars," *Science Advances* 4, no. 4 (2018): eaar6887; and Ryuki Hyodo et al., "On the Impact Origin of Phobos and Deimos," *Astrophysical Journal* 860 (2018).

17. A possible interconnected ocean or aquifer on early Mars, or extensive partial melting of the crust, would greatly enhance tidal dissipation in a nearby satellite.

18. First it will be tidally disrupted into a ring around Mars—that would be a spectacular sight for backyard astronomers, and Mars would triple in apparent brightness.

19. French planetary scientist Pascal Rosenblatt proposed in 2016 that a Vesta-sized proto-Phobos could fling other moons into higher orbits when it spiraled in. Subsequently, David Minton and Andrew Hesselbrock proposed that the timing need not be special, and envisioned a

"Mars ring cycle" where a large satellite spiraled in, broke up, threw a few 100-kilometer moonlets (say) into more distant orbits, which spiraled in and broke up, making Phobos-sized moonlets, and so on, so that 40 million years from now there will be some kilometer-sized nuggets and we will once more be wondering, "Why are we so lucky to see just these last two nuggets?" A. Hesselbrock and D. Minton, "An Ongoing Satellite-Ring Cycle of Mars and the Origins of Phobos and Deimos," *Nature Geoscience* 10, no. 4 (2017): 266–69.

20. Mars could well have changed its spin pole since then; the paleo-equatorial band can have been any great circle around present-day Mars, going through Borealis (if that was the impact event responsible).

21. A modern and detailed summary is provided by planetary geologist Edwin Kite in "Geological Constraints on Early Mars Climate," *Space Science Reviews* 215 (2019).

22. Here's another example of attrition bias. I came home one day to a house that had thousands of flies. Something had died and hatched . . . I spent two days killing them, vacuuming them from windowsills, swatting them, blowing them out the door with fans. In the end, there were eight final flies. Black little demons, they were smarter than the rest and had greater reserves and excellent strategies. When attacked by the vacuum hose, they would fly up and land on my hands, the only place I couldn't vacuum. How did they know? I spent three days killing the final eight. Two probably escaped, to breed a population of superflies.

23. Calculations by Robert Haberle and his group at NASA Ames Research Center in California, among others, showed the climate forcing of CO_2. Calculations by Jim Kasting at Penn State University rained on the parade, however, showing that CO_2 clouds would form, reducing the greenhouse effect.

24. Asteroid Apophis will come a tenth of a lunar distance from the Earth at 21:46 GMT on April 13, 2029. Now we know it won't hit us, but the encounter will perturb its orbit in a strongly nonlinear way; that is to say, a tiny change in its close approach will get magnified into a large change in its continuing orbit. So until the encounter is measured, we won't know where it's going next, except in a general sense. Sure, it may come back again to hit the Earth, but in fact there's greater probability that a random NEO that size or larger will hit the Earth in the meantime than for Apophis to come back. So there's nothing on the radar yet. But only about half the asteroids the size of Apophis have been discovered—a number that's rapidly closing thanks largely to NASA efforts directed by the U.S. Congress. Eventually the only unknown hazards will be asteroids smaller than a few hundred meters, and com-

ets. ("Comets are like cats. They have tails, and *they* do precisely what they want," to quote comet-astronomer David Levy, co-discoverer of comet Shoemaker-Levy 9.)

25. The first recognized impact craters included those with significant economic potential. Meteor Crater was bought for its nickel-iron mining rights. The diamond mines at Popigai were run as gulags under Stalin. The continental shelf off the coast of Chicxulub was recognized by PEMEX geologists as a possible impact structure, and they held it for years as an industrial secret.

26. The IDPs are further subdivided, and include pre-solar grains, dust grains with completely alien isotopic signatures that were around before the Sun was formed.

27. About 6 tons of the influx of cosmic materials per year comes in the form of micrometeorites, with about two distinct little meteorites landing per meter of land surface per year, according to a rooftop study that was spearheaded by Oslo jazz musician Jon Larsen in 2017. M. J. Genge et al., "An Urban Collection of Modern-Day Large Micrometeorites: Evidence for Variations in the Extraterrestrial Dust Flux Through the Quaternary," *Geology* 45, no. 2 (2017): 119–22.

28. Undifferentiated does not mean "never-melted." In close to zero gravity of a planetesimal, internal fluid stresses are orders of magnitude greater, as are magnetic and thermal stresses, so gravity won't simply pull together a core until the body gets to be tens of kilometers in diameter or larger, even if completely melted, I should think.

29. K is the symbol for the Cretaceous (from the German *Kreide,* or chalk), which was the last period of the Mesozoic era, and T is the Tertiary, the first period of the Cenozoic era, except that early Tertiary is now called the Paleogene, so K/T is a common anachronism and K/Pg is the more proper term. You can also call it the end-Mesozoic extinction, or end-Cretaceous, but "the K/T impact" is firmly adopted into the common parlance.

30. Forming kilometers deep, the seafloor sediments become a faithful record of what is going on at the surface. Whatever happens up on land is usually eroded away, so evidence is harder to come by.

31. "Snowball earth" has happened at least a few times in the Earth's geologic past, where the entire planet became white with ice and snow except for some volcanoes peeking up. The state of snowball earth leads to a distinctive rock record, marked by changes in oxygen and carbon isotopes, and terminated by abrupt carbonate deposition. When the oceans open up again, seawater can again dissolve the CO_2, and there is a precipitation of cap carbonates—layers that cap the snowball shutdown of sedimentation. These signs are not present at the K/T.

32. "Arrakis would be a place so different from Caladan that Paul's mind whirled with the new knowledge. *Arrakis—Dune—Desert Planet.*"— *Dune,* Frank Herbert (1965).

33. The equivalent today would be the "isotopic crisis" of lunar rocks being indistinguishable from Earth rocks.

34. This is the "broiled alive" hypothesis of Purdue University geophysicist Jay Melosh.

35. The first identified fragments of the K/T impactor were discovered and confirmed by cosmochemist Frank Kyte of UCLA, who also discovered the pieces of the Eltanin meteorite described earlier. It seems strange that fragments of *anything* can survive after impacting Earth's crust at 20 kilometers per second. But if you have an asteroid or comet impacting a planet, some small fraction of its volume will survive, little sectors where strong shocks are avoided just simply by luck. An impacting asteroid is like a "lithobraking" spacecraft, slowing itself down by ablating, while shielding some interior bits.

36. The GRAIL mission was actually two spacecraft, dubbed Ebb and Flow by the team led by MIT geophysicist Maria Zuber, that followed each other in close orbit about the Moon. The distance between them would stretch and shrink as they orbited, because the Moon's gravity field is not perfectly circular. By measuring this stretch extremely precisely, the team was able to derive a detailed gravity model of the Moon.

37. It extends from Aitken Crater, which sits at the northern extent of the basin, to the South Pole near its southern extent. So it is not at the South Pole, and it doesn't have a real name; for now we're stuck saying SPA.

38. Martin Jutzi and Erik Asphaug, "The Shape and Structure of Cometary Nuclei as a Result of Low-Velocity Accretion," *Science* 348, no. 6241 (2015): 1355–58.

39. The idea of cometesimal accretion originated ten years earlier, at the Planetary Science Institute, also in Tucson, especially the theoretical work by Stuart Weidenschilling, and the physical concept of primordial rubble piles advocated by Paul Weissman and others in the aftermath of the spectacular Giotto flyby of comet 1P/Halley in 1986.

40. Instead of theorizing, we can answer these questions directly, using radar tomography to make a high-definition medical-type scan. Mike Belton proposed the mission concept Deep Interior to NASA in 2004, and then I proposed it in 2009 and 2014 as Comet Radar Explorer (CORE). It uses 3D-imaging technology (related to seismology or ultrasound or CT scan) to see what's inside, in high-resolution detail. The proposal has made it twice to Category 2, which means it is "selectable," but as they say, that and a dollar will buy you a cup of coffee! A

high-definition image of the 3D structure of a comet or other primitive body will answer a number of the open questions in this book.

41. The key event of Deep Impact was to fire a 300-kilogram copper bullet into the surface. The impact was less informative than had been hoped, because the spacecraft was unable to see through the dense dust plume that it had just made. It produced a plume of sunlit dust, and by the time that had cleared away, the spacecraft was 1,000 kilometers downrange.

42. Modeling a small giant impact requires a small computer time step; thus many more cycles are needed to advance the simulation to completion. And now, with all this new physics, each time step has to calculate not only pressure, gravity, and temperature, but also damage, compaction, and the stress tensor. Pressure is a *scalar* (that is, just a value) that measures the magnitude of the stress tensor. The stress tensor has things in it "off-axis" that allow solids to have strength. Resistance to shear is resistance to a movement in the y direction, on a plane parallel to the x direction (resistance to slippage, that is). That shear stress is called s_{yx}, and so on for other planes. So there are nine components of stress in a solid (s_{xx}, s_{xy}, s_{xz}, s_{yx}, s_{yy}, s_{yz}, s_{zx}, s_{zy}, s_{zz}). The pressure is defined as all the stresses applied onto the same faces (that is, anything but shear), so $P=(s_{xx}+s_{yy}+s_{zz})/3$. And then the material's response to stress and pressure has to be defined—the *rheology* and the *equation of state*. The math is all basic algebra, but they are approximations, and there is an increasing possibility of bugs in the code, and the computations get time-consuming.

43. The barbaric practice where you don't get to eat lunch, but your audience eats while you present your research.

44. The talk, by Ian Garrick-Bethell, was on his idea that the early Moon solidified when it was captured into a 3:2 highly eccentric spin-orbit resonance, one that would have made it sufficiently out of round. The problem is, this resonance is hard to get into and out of.

45. Some call the Moon's farside the dark side, and it is after all in Earth's radio shadow, but day and night happen there every 28 days just like anywhere else on the Moon.

46. Lunar Orbiter images have been lovingly scanned and reprocessed by the United States Geological Survey (USGS) and the Lunar and Planetary Institute, and are available here: https://www.lpi.usra.edu/resources/lunarorbiter/.

47. The 60-megaton "Tsar Bomba" was exploded at high altitude in 1961, where 1 megaton is the equivalent explosion energy of 1 million tons of TNT. The Hiroshima explosion was 18 kilotons, or 0.018 megatons. Tsar Bomba was the equivalent of a few thousand Hiroshima explosions, and Bruno's formation was the equivalent of a few thousand of

those explosions. Had Tsar Bomba been a buried explosion instead of a high-altitude airburst, the result would have been a modern twin to Meteor Crater.

48. Verses 54:1–2 of the Holy Quran: "The Hour (of Judgment) is nigh, and the Moon is cleft asunder."

49. This was demonstrated by Paul Withers, "Meteor Storm Evidence Against the Recent Formation of Lunar Crater Giordano Bruno," *Meteoritics & Planetary Science* 36, no. 4 (2001): 525–29.

50. Note the caveat, however, that a crater the size of Bruno produces ejecta that orbits the Earth for some time and then comes back to hit the crater after it has finished forming. These can appear to be random asteroids, but are actually a sweepup pulse that gives the *impression* of age. So Bruno could be much younger than a million years, although still not from historic times.

51. If you are lucky enough to find yourself in the path of totality, look around for aliens or time travelers in disguise. Perhaps they will be abiding by the Prime Directive, which in the television series *Star Trek* prohibits Federation Starfleet from interfering with the natural internal development of pre-warp-drive civilizations. Please take them to our leader.

52. My favorite high-definition eclipse video in real time is by JunHo Oh and YoungSam Choi, taken from Warm Springs Indian Reservation in 2017: https://vimeo.com/231484786.

53. The black rectangular obelisk measures 1 × 4 × 9, the ratio of perfect squares. It appears one night in the desert grotto where Moon-Watcher's tribe slept, the Dawn of Man. The hirsute hominids pranced about it and exulted. The 1968 movie adaptation by Stanley Kubrick takes the shocking end of the book and turns it into an existential multicolored ghost ride to the surface of Europa, a psychedelic flip side to the immensely popular human spaceflight program at that time, and a distraction from racial and political assassinations and the Vietnam War.

54. You can look at the Sun with binocs *only* during the totality of the eclipse. If you look at the Sun with binocs during anything other than complete totality, you will go blind. You can buy eclipse lens covers for your binocs, or, as I recommend, special solar binocs with only one purpose, to stare at the Sun!

55. "Because I had only my writing brush and ink slab to converse with, I call it Brush Talks." *Complete Dictionary of Scientific Biography* (New York: Charles Scribner's Sons, 2008).

56. In one of his many stories, he wrote that "a man of Zezhou was digging a well in his garden, and unearthed something shaped like a squirming serpent, or dragon. He was so frightened by it that he dared not touch it, but after some time, seeing that it did not move, he examined it

and found it to be stone. The ignorant country people smashed it . . ." He wrote that this is an example, one of many, of sea creatures whose skeletons were been buried long ago and replaced by rock. He also noted petrified bamboo in regions where they could not grow in the present climate, and concluded that climate itself is subject to change. Joseph Needham, trans., *Science and Civilisation in China*, Volume 3, *Mathematics and the Sciences of the Heavens and the Earth* (Taipei: Caves Books, 1985).

57. From Shen Kuo, *Dream Pool Essays* (1088), as translated by Joseph Needham and Ling Wang in *Mathematics and the Sciences of the Heavens and the Earth* (Cambridge: Cambridge University Press, 1959). *Dream Pool Essays* is another name for *Brush Talks from Dream Brook*.

58. This interpretation is made by John S. Lewis in his book on near-Earth objects, *Rain of Iron and Ice*, rev. ed. (New York: Basic Books, 1997).

59. For a good scientific review of the Hadean landscape, to the extent that there was one, I recommend Kevin Zahnle et al., "Emergence of a Habitable Planet," *Space Science Reviews* 129 (2007): 35–78.

60. Matija Ćuk and Sarah Stewart, "Making the Moon from a Fast-Spinning Earth: A Giant Impact Followed by Resonant Despinning," *Science* 338, no. 611 (2012): 1047–52. This paper is controversial for suggesting that angular momentum is not necessarily conserved, so that the early Earth might have been spinning near the breakup limit, 2.4 hours as they proposed—not far from what Darwin had required.

61. English astronomers Alan Jackson and Mark Wyatt have studied the idea that the Earth and Moon suffered a tremendous bombardment by material coming back from the original giant impact—more than a lunar mass of material overall.

62. Hence the mission name Lucy, after the skeleton of *Australopithecus afarensis* discovered in Ethiopia in 1974, from around the time of the earliest stone tools, which in turn was named after the Beatles song "Lucy in the Sky with Diamonds," which seems appropriate to the mission as well.

63. There could have been two Trojan moons more comparable in size, with only one of them colliding with the Moon and the other colliding with Earth. This is dynamically rather plausible, although the scenario of two unequal Trojan moons, one maybe 30 to 100 times more massive than the other, is consistent with the record of Trojan moons at Saturn. In any case a mini-Trojan might not survive long enough to contribute to the Moon's geologic record; if it did, there might be a smaller splat somewhere on the Moon.

64. When the Moon was orbiting close to the Earth, the bombarding projectiles were accelerated by the gravity of Earth. Anything that hit the Moon was going at many times its escape velocity, so it was a hazardous

time for the Moon until it evolved tidally to be tens of Earth radii away. The impacts kept coming, and at very high velocity compared to the Moon's escape velocity, which is only 2.4 kilometers per second, so the result would have been a sandblasting effect, even causing the Moon to lose hundreds of kilometers of its original radius. This has yet to be factored into any theories for lunar crust formation.

65. To be published, a scientific result has to make it through *peer review*. That is, the article has to be approved by a peer in the field, or (for the best journals) two peers plus an editor. Peer review is fundamental to science, and it is not at all mysterious why it should work. One's scientific reputation is analogous to being a host for Airbnb. Your five-star rating (the reliability and value of your research) means you win more grants and participate in fun stuff like missions. And you will only stay in a house (rely on someone's paper) that has a good rating. You publish good papers in good journals, and strive for a reputation as a fair but critical referee. Peer review is not a guarantee that the publication is correct, just that it is of scientific value.

66. The Moon is not flat at 1,300-kilometer scales, which is why we use crater scaling only as a guide. Crater scaling laws are developed for what is described as a *half-space*—that is, a geometry that is infinite in all directions beneath a flat surface. For Tycho-sized and smaller impact craters on the Moon, it is effectively a half-space.

67. China just landed there for the first time with Chang'e-4 and is gearing up for a sample return with Chang'e-5.

68. A shock wave happens when the impact velocity ramps up; the energy of a wave overtakes itself, like a boat riding on top of its own wake. A strong shock wave is created when a projectile impacts a target at a velocity faster than the speed of sound; the energy can't get out of the way as fast as it's coming in. For terrestrial planets, this is about 5 kilometers per second; for icy planets, it's about 3 kilometers per second. Realistically, pockets of brine and cracks in the rocks and layers will cause an impact to "shock up" at lower velocities, leading to crushing and damage and frictional heating. The physics, thermodynamics, and geology of collisions is best described in the textbook by H. Jay Melosh, *Impact Cratering: A Geologic Process* (New York: Oxford University Press, 1989).

CHAPTER 7: A BILLION EARTHS

1. Most of the mass is in the largest bodies, but most of the *surface area* is in the smallest bodies, the dust. That is why there is a zodiacal light, sometimes, the glow of the ground-down dust from the Main Belt. You

can't see the asteroids with the naked eye, but you can sometimes see the dust, although it has negligible mass.

2. Not really ices. Cometary activity is thought to derive in part from amorphous solids that never crystallized into ices. But they're cold like ice, and made of the same stuff (e.g., H_2O, CH_4) as ice, just not in crystalline form.

3. The Psyche mission goes to asteroid Psyche, arriving in 2028. The Lucy mission goes to six Jupiter Trojan asteroids, arriving between the years 2025 and 2033.

4. Tungsten-hafnium dating is one of the most fascinating of the radiometric clocks. Hafnium is an element that tends to become part of the rocky mantle (a *lithophile*). One of the isotopes of hafnium, ^{182}Hf, is radioactive and spontaneously breaks down into stable tungsten, ^{182}W (the metal used to be called wolfram). This transformation has a half-life of 9 million years. Whenever a planet melts and differentiates into a mantle and a core, hafnium will stay in the rocks, and tungsten will go to the core, with the other metals. If a planet differentiates within a few million years and *then* solidifies, its rocks will still have "live" ^{182}Hf locked in their crystals. When these break down inside the solid crystals, they leave an accumulating amount of tungsten that scientists (billions of years later) read as a clock. Hafnium is called an *extinct radionuclide* because although the clock is long dead—all the radioactive Hf has decayed—the hour hand points to a certain time.

5. A review of the timing of mantle and ocean evolution is provided by geochemist Linda Elkins-Tanton, "Formation of Early Water Oceans on Rocky Planets," *Astrophysics and Space Science* 332, no. 2 (April 2011): 359–64.

6. Viranga Perera et al., "Effect of Re-impacting Debris on the Solidification of the Lunar Magma Ocean," *Journal of Geophysical Research* 123 (2018).

7. The reduction-oxidation state ("redox") describes the amount of free hydrogen, or free oxygen, relative to water, H_2O. Oxidizing conditions means there is oxygen around, looking for a partner. The Earth's surface is in an oxidizing state, which is why everything rusts, but this is also why you can breathe. Jupiter is a highly reducing atmosphere, in that any free oxygen would instantly find hydrogen.

8. Jeremy Bellucci et al., "Terrestrial-Like Zircon in a Clast from an Apollo 14 Breccia," *Earth and Planetary Science Letters* 510 (2019): 173–85.

9. John Armstrong, Llyd Wells, and Guillermo Gonzalez, "Rummaging Through Earth's Attic for Remains of Ancient Life," *Icarus* 160, no. 1 (2002): 183–96.

10. Willy Benz wrote the original smoothed particle hydrodynamics (SPH)

code for simulating giant impacts in planetary science. The use of *hy-drocodes* like SPH has added a numerical laboratory to modern plane-tary physics, allowing us to study problems well outside the laboratory by coding up a way of integrating the observed physical equations, starting with the fundamental constraints, the conservation of mass, momentum, and energy.

11. Canup and Asphaug, "Origin of the Moon": 708–12.
12. TRAPPIST stands for Transiting Planets and Planetesimals Small Telescope, an acronym referencing the famed brewer monks of Bel-gium. Naming of planets around stars is not yet as systematic as the naming of comets and asteroids. There is "Tabby's star," named after astronomer Tabetha Boyajian, who spent many years monitoring this crazy system that brightens and dims almost at random and is probably orbited by dynamically unstable debris. If the star has a name, such as 51 Pegasi, then its first-discovered planet is usually given a suffix, in this case 51 Peg b (the first planet ever discovered around a main sequence star, "a" being the star itself).
13. Audiophiles will recognize this as SNR, the signal-to-noise ratio.
14. On Mercury there are regions near its poles, tens of meters under the surface, where mats of liquid water could be present in small quantities and stable for a long time.
15. Ludwig Wittgenstein, *Tractatus Logico-Philosophicus* (1918). "Wovon man nicht sprechen kann, darüber muß man schweigen."
16. Peter Ward and Donald Brownlee, *Rare Earth: Why Complex Life Is Uncommon in the Universe* (New York: Copernicus Books, 2000). Despite there being "billions and billions" of planets in the cosmos, they argue that the conditions for complex life are exceedingly rare, although microbial life might be common.
17. One of the criticisms of *Rare Earth* is that it is Earth-centric; it is one thing to argue that Earthlike conditions are rare (plate tectonics, a Moon-like moon, and so on), but that doesn't mean that all of these conditions are necessary for complex life. See *Life Everywhere* by as-tronomer David Darling (New York: Basic Books, 2002).

CONCLUSION

1. This has sometimes been NASA's small-missions mantra: faster, bet-ter, cheaper. The reply is usually "choose two."
2. Named for the British polar explorer Ernest Shackleton.

INDEX